LARVAE AND EVOLUTION

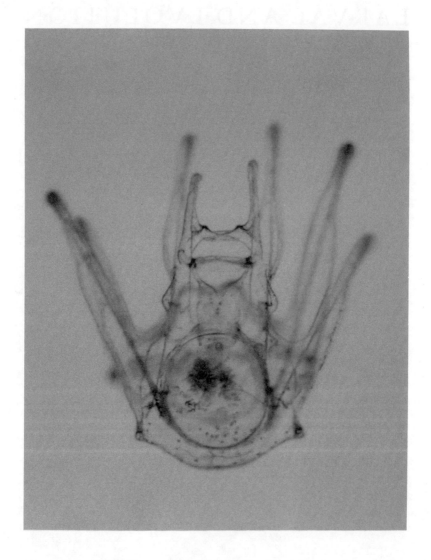

Frontispiece. A hybrid larva, 29 days old, developed from an egg of the sea squirt *Ascidia mentula* (phylum Chordata, subphylum Urochordata) fertilized with sperm of the sea-urchin *Echinus esculentus* (phylum Echinodermata).

LARVAE *and* EVOLUTION

Toward a New Zoology

DONALD I. WILLIAMSON

Port Erin Marine Laboratory

University of Liverpool

Foreword by Lynn Margulis and Alfred I. Tauber

University of Massachusetts

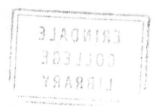

CHAPMAN AND HALL

New York   London

First published in 1992 by
Chapman and Hall
an imprint of
Routledge, Chapman & Hall, Inc.
29 West 35 Street
New York, NY 10001-2291

Published in Great Britain by
Chapman and Hall
2-6 Boundary Row
London EC1 8HN

Library of Congress Cataloging in Publication Data

Williamson, D. I. (Donald Irving)
  Larvae and evolution / D.I. Williamson.
    p.   cm.
  Includes bibliographical references and index.
  ISBN 0-412-03081-0
  1. Invertebrates—Evolution. 2. Invertebrates—Larvae. 3. Evolution
(Biology) 4. Larvae. I. Title.
  QL362.75.W54 1992
  592'.038—dc20   91-25672
                                                          CIP

British Library Cataloguing in Publication Data also available.

To my wife, Enid,
who not only gave moral support
but also drew many of the figures

Contents

III. SOLUTIONS

IV. CONCLUSIONS

Foreword

In the spring of 1988, Lynn Margulis received an unsolicited letter from the Isle of Man which began: "I am 66 years old, from a family whose members are short-lived and thus I'm on a straight-line course for posthumous recognition." The letter from Don Williamson went on to suggest that his and Lynn's scientific careers have aspects in common: Williamson briefly presented his theory as you have it here in this book and his litany of publication woes. The original paper, "Incongruous larvae and the origin of some invertebrate life-histories," published in *Progress in Oceanography* (19:87–116, 1988), had been rejected by seven other journals, before he found an intrepid editor, who commented:

> It is sometimes stated that papers involving ideas are more difficult to get published than those that follow the conventional wisdom of the time. Iconoclasts are rarely popular with their peers. Here is a paper that I am including in Progress in Oceanography as something of an editorial indulgence, but which I am confident that in time the decision will be vindicated.
> . . . Darwin would have probably had less trouble submitting a draft of the *Origin of Species* to the Bishop of

Oxford. From the content of some of the reviewers' comments, I am convinced that prejudice has blinded them to the logic of the arguments presented by Dr. Williamson.

. . . In this paper Williamson takes a fresh look at what is known about several of the larval forms familiar to planktonologists. With a simple, but to some unacceptable, conceptual leap, he presents an explanation for the evolution of certain of the recent larval forms which seems both rational and feasible. Although conclusive proof will never be forthcoming, many of the elements of Williamson's theory are testable. Biochemical techniques can be used to demonstrate just how close some of the proposed relationships really are. Similarly, the possibility that the fusion of disparate genetic material can occur and hence might occur under natural conditions, can be further explored. Already for many cell biologists the theory will not appear so very far-fetched, because the origins of organelles which are cytoplasmically inherited, such as mitochondria and chloroplasts, are thought to lie in the fusion of two different taxa to form a single entity. The original association is thought to have come about through a symbiotic or parasitic relationship. There are many examples of associations which may be analogous to the intermediate steps.

. . . The fact that genetic engineering techniques are based on overriding the (immune) responses, shows that even the more elaborate systems of higher Metazoa can be overcome. . . . There will still be those who believe that this paper "pollutes the scientific literature," but I remain unapologetic. The essence of science is the development and testing of ideas and theories, and this paper will certainly stimulate debate and experiment.

We concur—Don Williamson deserves to be heard.

Williamson requested Lynn Margulis's aid in bringing his ideas to the attention of the serious scientific public: to that time he had published over 70 papers of related interest in highly specialized journals, but regarding the theory per se, he had written only a batch of incomplete manuscripts and unpublished book material. Lynn read his claims 1) of the fertility of nonreciprocal interphyletic crosses (between sea urchin sperm and tunicate eggs that form pluteus larvae), and 2) of the obsolescence of Haeckelian family trees for marine animal phyla that should be replaced with an anastomosing phylogenetic diagram. The reason that so many classes of animals have similar trochophore, veliger or pluteus larvae has nothing to do with ancestry, argued Williamson, but everything to do with the ability of these larvae to disperse the developing animals. The Williamson theory, although focused on larvae, addresses the general problem of parallelism, where the same character re-appears in distantly related taxa, while it is absent in more closely related groups (see e.g. Ronald Sluyś discussion concerning phylogenetic analysis, Syst. Zool. 38:350–370, 1989). Notwithstanding neo-Darwinian explanations, Williamson has offered a radical "one-step" alternative and the initial experimental confirmation of his theory.

Although they seemed idiosyncratic, the ideas were fascinating. Lynn suggested he write his thesis in shortened form for the *Proceedings of the National Academy of Sciences.* Don did this with alacrity. Since the first four reviews returned were inconclusive, still others were sought. One reviewer refused to read on, branding the ideas as high-school level rubbish. Another reviewer thought the fossil record evidence was fully consistent with Williamson's thesis, but that it was difficult to evaluate his statements about embryology and development. A third reviewer felt that Margulis' reputation as a scientist would be jeopardized by any further relationship with Williamson's work. Still another reviewer felt that Williamson's scenario deserved a thorough "dispersion" of its own. Margulis agreed with nearly every reviewer who requested closer photographic documentation of the live organisms discussed. A later reviewer insisted that molecular data verify Williamson's claims; all readers agree that his concepts lead to testable verification by these methods.

While Williamson's work was under review, Fred Tauber began arranging a symposium on "the self." When Fred requested names of appropriate biologists who had seriously considered the issue of what conferred selfhood for animals other than humans, Lynn immediately nominated Williamson and the controversy surrounding his fascinating work. Could Williamson be very clever but deceptive? Was Williamson on to one of the most interesting ideas in a century involving marine zoology? Could Williamson really convince a sophisticated audience that his theory was based on adequate observation and sound scholarship? Before the Symposium, during a trip to the British Isles, Fred Tauber detoured to Port Erin and wrote Margulis from Williamson's marine biology laboratory. Fred was convinced that Williamson was an erudite and careful scientist whose originality needed airing and appropriate criticism. Although he had little in the way of modern technology, Williamson had performed decades of careful work with equipment at least as good as that of Ernst Haeckel. After all, Williamson was certainly well-known in Korea and Japan for his research on shrimp and other crustaceans, more so than in his native land or other English-speaking communities.

In April 1990, Williamson presented his theory at the Boston University Colloquium for Philosophy of Science and at a seminar at the Marine Biological Laboratory at Woods Hole. (The conference was published: *Organism and the Origins of Self*, Kluwer Academic Publishers, 1991. A synopsis of *Larvae and Evolution* may be found there—"Sequential Chimeras," pp. 299–336.) At the end of his talk to some 30 people in the cavernous B.U. Law auditorium that holds 500, Fred rose up and assessed his audience—including some apoplectic sea urchinologists: either we were all privy to an historical moment (as the Haeckel phylogeny itself went extinct to be replaced by that drawn by Williamson) or else Williamson would be recorded as an ordinarily competent and diligent zoologist, who lived with animals by the seaside for 30 years, and whose mistake should be forgiven in the light of his standard contributions.

Our editor at Routledge, Chapman and Hall, a man fascinated by good ideas, Mr. Gregory Payne, took the leap. He agreed that whatever the fate of Williamson's original concepts of "borrowed" larvae and anomalous life histories, these hypotheses are subject to scientific inquiry. Williamson's theory deserved the spotlight of scientific scrutiny and the courtesy of a proper hearing. Greg, accordingly, brought this manuscript from its rough-and-ready state to the proper expository form in which you now find it.

We all believe that scientists with unpopular ideas need not suffer suppression. We suspect that many marine zoologists have noted similar phenomena as those unified by Williamson's theory. We entirely support Williamson's privilege to bring this case to his peers. We encourage you to read this fascinating text and consider the author's documentation. Better yet, *test* the theory. Optimization of the cross-fertilization experiments has not been attempted, nor such procedures as micropuncture to bypass early fertilization steps to maximize the yield of hybrids and increase the probability of observing metamorphosis. Biochemical comparison of selected enzymes and the use of immunological probes would be important confirmations of the morphological data. In addition, at the very least, 1) a swimming ciliated blastula inside an ascidian chorion or hatching from it must be shown, 2) karyotype analysis for determining nuclear origins must be performed, as well as 3) delineation of whether or not the hybrids have ascidian mitochondrial genes. Although these experiments should be performed, Don Williamson should not be chastized for the fact that no one has undertaken this work. He has functioned alone in a 19th century mode for too long. He has lacked access to modern methodology. As a retired Reader (Professor) at the University of Liverpool he has no resources to pursue the scientific problem with current techniques. His theory is based solely on observation, intuition, a broad grasp of phylogenetic relationships, and the tantalyzing first experiments that appear to support his argument. There is of course a poignancy in his pleas for colleagues to apply modern techniques to study the chimera theory. Why, we wonder, should Ernst Haeckel with his century-old data be more correct than Donald Williamson when it comes to diagramatic presentation of the relationships between the major groups of marine animals? The question deserves re-examination.

Lynn Margulis, Ph.D.
Distinguished University Professor
University of Massachusetts, Amherst

Alfred I. Tauber, M.D.
Professor of Medicine
Boston University School of Medicine

April, 1992

Preface

When Charles Darwin published his *Origin of Species* in 1859 it was considered heretical by most people. Today the situation is reversed, and, in biological circles at least, it is considered heretical not to agree with Darwin. I am perpetrating only a minor evolutionary heresy, for, although I disagree with him on one important point, I do not dispute either that evolution has taken place or that natural selection has played a fundamental part in it. One of Darwin's assumptions was that the ancestry of all living and extinct organisms can be depicted in a tree-like diagram, the branches of which never rejoin. The evolution of some animals and their larvae, however, may be explained on the assumption that occasionally some branches have interacted with others with profound repercussions, and these reactions have not been confined to nearby branches. This is the theme of this book. It points out that some very diverse groups of animals have similar types of larvae whereas some dissimilar larve give rise to similar adults, and, to explain these and related phenomena, a new hypothesis is proposed. This postulates occasional transfers of genetic material between animals that are not closely related. It may be regarded as an addition to Darwin's theory or a modification of it, but in so describing it

I have no delusions that my work is comparable to his or I to him.

I am very grateful for many valuable suggestions for the improvement of the manuscript by Lynn Margulis, Frederick R. Schram, and the editor. The deficiencies that remain, however, are mine, not theirs.

I
Introduction

This is a book about one aspect of the links between the development of animals and their evolutionary history, between their ontogeny and their phylogeny. The very mention of "ontogeny and phylogeny" will remind many biologists of the erudite book of that title by Stephen Jay Gould (1977) or the earlier works *Embryology and Evolution* and *Embryos and Ancestors* by Gavin De Beer (1930, 1940), but there is virtually no overlap between these books and the present work. Gould and De Beer both explored the extent to which animals recapitulate their previous evolution during their development and the evolutionary effects of acceleration and retardation of development. I wish to explore an entirely different link between the way many animals develop and the way they have evolved. Evolutionary theory, as it is today generally accepted, postulates that species evolved from other species in monophyletic lines of descent by the accumulation of gradual, heritable changes. I agree that this accounts for much of evolution, but I claim that this theory alone is not adequate to explain why some animals have the embryos and larvae they do and the life histories they do. In suggesting an additional hypothesis to explain these anomalies, I must make it clear that my proposal is an addition to accepted theory, not a replacement for it. My explanations all assume that conventional, Darwinian evolution has taken place and is continuing to take place. Those who wish to attack the entire theory of evolution are unlikely to find suitable ammunition in these pages, and they should certainly not include me as their ally.

The stages that develop from fertile eggs are frequently very different from the adult animals that develop from them, and, if they are sufficiently different, the young stages are called larvae. The larval and adult phases in development commonly follow totally different ways of life, and the processes that Darwin called "descent with modification" and "natural selection" are frequently adequate to explain the differences in form that adapt them to their different environments. Paradoxes exist, however, in the many examples of apparently closely related larvae giving rise to apparently distantly related adults, and the less common examples of apparently distantly related larvae giving rise to apparently closely related adults. I shall be concerned with both these categories.

Conventional explanations, where they exist, invoke either convergent evolution or larval conservatism. Although convergent evolution undoubtedly occurs, I think that it provides a very inadequate explanation of similar planktonic larvae of vastly different adults. Adult and larval forms sometimes evolve at different rates, but I question the more extreme cases that have been attributed to adult divergence and larval conservatism in the same lineage. My alternative theory, horizontal transfer of larval form, does not postulate that larvae have altered very little or become superficially more similar as their corresponding adults have evolved striking differences. Rather it proposes that, on occasion during evolutionary history, larval and embryonic forms that originally evolved in one lineage have later appeared in another, as if they had jumped from one branch of the phylogenetic tree to a distinct and sometimes distant one. The implied transfers of large amounts of genetic material are attributed to successful hybridizations between animals that are not closely related. This assumes that the genes specifying larval form act largely independently of those specifying adult form, but most animals keep their larval and adult morphologies quite separate, irrespective of how the larvae evolved. My proposals also affect methods of metamorphosis, for animals with new larvae from a remote group will have to invent new methods of reaching the next phase in development.

I suggest that larvae of sponge crabs (Dromioidea) resemble those of hermit crabs (Paguroidea) not because of convergent evolution of the larval forms but because an ancestor of all the modern sponge crabs acquired a larva from a hermit crab. The larvae of echinoderms resemble, to varying degrees, those of acorn worms (Enteropneusta) not because of common ancestry but because echinoderms had no planktonic larvae until one of them acquired larvae from an acorn worm. Derivatives of this larval form were then acquired successively by ancestors of all the echinoderms that today have larvae. Among these derivatives are the so-called pluteus larvae of brittle-stars and sea-urchins, the only larvae to have arms supported by calcareous rods. I hold that the similarity between the larvae of these two classes of echinoderms is evidence not of common ancestry but of larval transfer from a sea-urchin to a brittle-star long after the adult groups were established. I challenge the idea that the trochophore larvae of polychaete worms, echiurans, sipunculans, and molluscs and the broadly similar larvae of nemertines, bryozoans, and polyclad flatworms provide evidence of direct descent from an ancestor with a larva from which all these larval forms evolved. Others (e.g., Jägersten, 1972) have suggested that the polyclad larva was the starting point for the larval forms of all these very diverse groups, and I agree. I suggest, however, that a representative of each group acquired its larva by hybridization either directly with a polyclad or with another animal that had already acquired its larva, directly or indirectly, from the same source, and I further suggest that the groups that acquired new larvae did so well after the adult features of the phyla were established.

When an adult (or a second larva) is derived from only a

small part of the larval tissue and the remaining larval tissues are digested or discarded, I take this an indication that the larval form was originally acquired from a distantly related group, probably from another phylum. Examples of such extreme forms of metamorphosis are the change from an echinoderm larva to an adult or the change from a polychaete trochophore to a segmented nectochaete, which is a second larva. When a larva metamorphoses by differential growth of established tissues, I take this as an indication that the larva may have originated in the same lineage as the corresponding adult, as in the change from a tadpole to a frog. A comparatively smooth metamorphoses may occur, however, when a larva has been transferred from a related group, such as from a hermit crab to a sponge crab, both of which are decapod crustaceans.

A species, as Darwin frequently pointed out, comprises all stages or phases in development. In a multicellular animal, these phases may include the egg, the embryo that develops within the egg, the hatched larva, the juvenile that metamorphoses from the larva, and the reproductive adult. For most of my career as a biologist I assumed, as Darwin did, that a species can have only one line of descent, from species to species, and that clues to that lineage provided by the egg, embryo, and larva were just as valid as those provided by the juvenile and adult. The theories that I am now putting forward imply that this assumption is frequently invalid, as the phylogenetic origins of the embryo and larva may differ from that of the adult. For example, the embryo and larva, which are today part of the life history of species A, may have been acquired by genetic transfer from species B, A and B being remotely related in some cases, more closely in others. A and B would have continued as distinct species after the transfer, but the eggs of the specimen of species A affected by the transfer would have hatched as the same type of larva as species B. In some instances, some of these larvae would have metamorphosed into juveniles of species A, and any progeny of the resulting adults would have repeated the same life history, with B-type larvae metamorphosing into A-type juveniles.

At first sight this may seem a highly improbable thesis, but it provides a tenable explanation of the facts. When it is realized how many paradoxes it can solve, how much evidence is consistent with it (and none inconsistent), and that the mechanisms proposed are testable by a variety of techniques, then I hope that the search for new evidence will continue, the relevant tests will be carried out, and new tests will be devised.

Evolution depends on genes, factors that control heredity that are normally passed unchanged from one generation to the next but that can occasionally change either their chemical composition or their physical relationship to other genes to modify their expression in the organism. Darwin published his theory—that organisms have evolved by descent with modification and by natural selection working on the modifications—long before others demonstrated the existence of genes or genetic mutation and recombination. This gives me some comfort in trying to add a postscript to his theory, for my own hypothesis of the transfer of larval form depends on genetic mechanisms that are imperfectly understood but that can be inferred. It may be many years before these mechanisms are fully resolved, but there is a good precedent for publishing the evolutionary theory first and allowing the genetics to catch up later.

In the chapters that follow, after outlining the background to the problem, I introduce the concept of horizontal larval transfer with two examples in which the donor and receiver groups are both decapod crustaceans. These cases illustrate that the phenomenon may affect only a small minority of the group concerned and does not necessarily involve an extreme form of metamorphosis. By contrast, the echinoderms, which are considered next, provide an example that I interpret as interphylar transfer, covering most of the extant members of the recipient phylum and involving drastic forms of metamorphosis. The way in which the phenomenon has affected different classes of echinoderms and some individual species is also explored, and the fossil record of the phylum is shown to be consistent with the new views being put forward. Those groups of animals that develop from trochophores and similar types of larvae provide examples attributable to transfers of related larvae to several remotely related phyla, and they illustrate other extreme forms of metamorphosis. These examples, which indicate the range of the phenomenon, are followed by an account of the suggested mechanism and experiments related to it. Fortuitously this sequence followed in the book corresponds roughly to the order in which the ideas occurred to me, not only the order of the groups regarded as having been affected by larval transfer but also the later hypothesis on the proposed method of genetic transfer.

My ideas on larvae and evolution are new, but the observations on which they are based are not, for these are scattered through the biological literature of over a century. The present book contains a fairly extensive bibliography, but additional references to publications immediately relevant to the theory are included in two articles by myself (Williamson, 1988a, b), and all the older literature is listed and summarised in the five volumes of The Invertebrates by Hyman (1940–59). Page references to The Origin of Species refer to the edition published in 1985 in Penguin Classics, which is a reprint of the first (1859) edition but with new pagination.

I
OVERVIEW

2
Larvae

Larvae of particular species and groups of animals are described in succeeding chapters, but here the subject is introduced by considering some general features of embryonic and larval development and defining relevant types of larvae.

A larva may be described as an immature phase, differing considerably from the adult, in the postembryonic development of an animal. To develop into the adult, the larva must metamorphose, i.e., change form, but there is no definition of the amount of change that constitutes metamorphosis. A distinction may be drawn, however, between development of the juvenile by relative growth of larval tissues and development in which the juvenile grows from a few discrete pockets or patches of tissue within the larva.

An animal starts its development in the egg, as an embryo, and the amount of development that takes place before hatching varies considerably from group to group, and frequently from species to species within the same group. For the moment, let us consider only those groups that are sufficiently distinct to be called phyla, the major groups (apart from superphyla) into which the animal kingdom is divided. All members of some phyla, such as the arrow worms (Chaetognatha), hatch in a form resembling miniature adults. They have no larvae, and development is said to be direct, although it would be more accurate to say that their postembryonic development is direct. There is no metamorphosis after hatching, but there is considerable change of form in the egg. Animals that never had a larval phase throughout their evolutionary history show primary direct development, whereas those without larvae but that are descended from species with larvae show secondary direct development. Jägersten (1972) believes that there are no true cases of primary direct development, but I recognize this category for the Chaetognatha and other animals that show no trace of a larval form at any time during their lives.

In contrast to the Chaetognatha, there are phyla, such as the Endoprocta, in which all known species have larvae, but in most phyla some members have larvae and others do not. Some animals, like the polychaete worm *Spirorbis,* have a larval life of only a few hours, whereas the European eel, *Anguilla anguilla,* takes 3 years to complete its larval development. The larvae of *Spirorbis* and *Anguilla* are very different, but there are many examples of animals with the same general type of larva covering the range from abbreviated to extended larval life. The range within shrimps of the genus *Pandalus,* for example, extends from 1 day to several months.

Most marine larvae are planktonic, with swimming powers sufficient to influence their depth but insufficient to have much direct effect on their horizontal distribution. A few are bottom living or parasitic. Terrestrial crustaceans with larvae migrate to the sea to release larvae, and a few freshwater shrimps migrate to estuaries for larval release. Many crustaceans, however, inhabit fresh water both as adults and larvae. Terrestrial insects with larvae release their eggs so that the larvae develop in fresh water, on land, or in or on the bodies of other animals.

Metazoans include all animals whose bodies are made up of more than one type of cell. They all go through a stage, before hatching, in which the dividing cells form a more or less hollow sphere, the blastula (examples are shown in Figs. 5.2a and 7.4a). The fluid-filled space within the blastula is the blastocoel. In many echinoderms the cells of the blastula are ciliated and the animal hatches in this stage (Fig. 5.2a), but in some echinoderms and most other animals embryonic development continues further. The stage that follows the blastula, whether embryonic or larval, is the gastrula, in which there is not only a spheroidal outer layer but also a tube-like inner layer of cells, the archenteron or primitive gut. The inner end of the archenteron is closed, and the opening at the outer end is called the blastopore (Fig. 5.2b). In the larval development of echinoderms, the body cavity or coelom starts to form from outgrowths of the archenteron, which soon form separate sacs (Fig. 5.2c,d). The blastopore becomes the anus, and the mouth develops as a new opening (Fig. 5.2e). This type of coelom, derived from the archenteron, is termed an enterocoel, and this type of mouth, which develops independently of the blastopore, is termed a deuterostome; the terms enterocoel and deuterostome are also applied to animals with these features. Apart from the echinoderms, the only other major group whose larvae are propelled by cilia and that develop by enterocoely and protostomy are the acorn worms (Enteropneusta). In early larvae of these two groups the cilia may be limited to a circumoral ring (Fig. 5.2e), but in later larvae there may be more than one band of cilia and the patterns are varied and frequently complex. The auricularia has a single, convoluted band of cilia (Fig. 5.3d). In the bipinnaria (Fig. 5.3a) and tornaria (Fig. 6.1c) there are two bands, both convoluted in the bipinnaria, the posterior band circular in the tornaria. A brachiolaria is a late bipinnaria with organs of attachment. The bands are drawn out into lobes in some bipinnarias (Figs. 7.2, 9.3c) and into slender arms supported by rods in pluteus larvae (Figs. 5.3b,c, 6.1, 7.3, 8.1, 8.2a–d, 9.3a,b). Doliolaria larvae are barrel shaped and have cilia in three to five separate, simple bands (Figs. 6e, 13e).

Trochophores and similar types of larvae occur in at least seven phyla. They swim with cilia, as do the larvae of echinoderms and acorn worms, but in the larvae now under consideration the cilia are grouped in a preoral ring and, usually, apical tufts. The ring of cilia is circular in the trochophore (Figs. 10.2a, 10.3a,b,d,e, 10.4a,c), forms four lobes in the pilidium (Fig. 11.1c) and Götte's larva, and eight or ten lobes in Müller's larva (Fig. 11.1a). The cyphonautes (Fig. 21f) are larvae with triangular, bivalved shell. Larvae of this group differ from the larvae of echinoderms and acorn worms in the method of formation of the mouth and, when present, the coelom. The mouth of trochophores and near-trochophores always forms before hatching, and it is a protostome, which means it develops directly from the blastopore. Not all larvae of this type metamorphose into animals with a coelom, but the coelom, when it occurs, usually develops after hatching, and it is a schizocoel. It does not originate from the archenteron but from splits in the connective tissue between the ectoderm and endoderm.

In some cases trochophores metamorphose to produce not a juvenile but a second larva. A trochophore larva is not segmented, coelomate, or shelled, but a polychaete trochophore may be succeeded by a segmented, coelomate nectochaete, a sipunculan trochophore by a coelomate pelagosphaera (Fig. 10.3f), and a molluscan trochophore by a shelled veliger (Fig. 10.4b,c). In all these cases the ciliated, planktonic second larva will undergo a second metamorphosis to produce the adult, which is usually benthic. The larva of some molluscs is neither a trochophore nor a veliger but a test-cell larva, also known as a pericalymma (Fig. 10.4d,e). This larva is covered with very large cells, apart from a posterior opening, and it has cilia grouped in an anterior tuft and one or more transverse rings. It metamorphoses directly to the juvenile.

Tadpole larvae occur only in the chordates (phylum Chordata). They swim not with cilia but with undulations of a muscular tail. Such larvae are comparatively well developed at hatching and have a brain and dorsal nerve cord, a heart, and a notochord supporting the tail. Tadpole larvae of ascidians and other urochordates are considered in Chapters 13 and 14 (Fig. 13.1).

Arthropods have no cilia at any time during their development. All postembryonic arthropods have chitinous cuticles, and appreciable growth is possible only when the cuticle is moulted. The development of external features is therefore divided into a series of intermoults, instars, or stages. Larvae of two groups of arthropods are relevant to the present work. A number of crustacean larvae are illustrated in Figures 3.2 and 4.1, and there are brief references to insect larvae and methods of metamorphosis. Several types of crustaceans hatch as nauplii, i.e., larvae with three pairs of functional appendages and usually a small, median eye. The appendages are all cephalic and will become the two pairs of antennae and the mandibles of the next phase in development. In some crustaceans the nauplius is an embryonic phase, and the animals hatch as a zoea. Larvae of this type swim with the outer branches (exopods) of some or all of their thoracic appendages. They typically have large, paired eyes and all or most of the adult complement of body segments. Types of crabs and hermit crabs are discussed in Chapter 4, and, after two or more zoeal stages, these crustaceans metamorphose to give a second type of planktonic larva, the megalopa. This phase swims with its abdominal pleopods for a period ranging from days to months, before it settles and metamorphoses into a juvenile.

All insect larvae have the same number of body segments as the adult. Some have appendages, others do not. Caterpillars have three pairs of thoracic walking appendages (legs) and a variable number of paired abdominal appendages (prolegs).

Blastula larvae are usually between 0.1 and 0.2 mm in diameter. Some species of trochophores and nauplii hatch at about 0.15 mm, but others are considerably bigger. Newly hatched crustacean zoeas range in length from about 1 to over 10 mm. Lecithotrophic larvae, i.e., those that feed only on yolk, are usually bigger at hatching than comparable planktotrophic larvae, larval life tends to be abbreviated, and there is little growth during larval development. The size of most planktotrophic larvae gives a rough clue to their age, and some that hatch at about 0.1 or 0.2 mm will grow to over 10 mm after a planktonic life of several months. Others, however, such as bivalve veligers, cease growing after a few days, and in such cases there is little indication whether they have spent days, weeks, or months as planktonic larvae.

3
The Issues in Context

Darwinian Evolution—Non-Darwinian Evolution—Genetic Engineering and Viability of 'Foreign' Genes—Classification and Phylogeny—Darwin's Example of Barnacle Larvae and Classification—Increased Knowledge of Marine Larvae Since Darwin.

The broader topics at issue in this book are evolution, "foreign" genes, classification, and larvae. It is instructive to call to mind, where applicable, Darwin's views on these subjects and briefly review progress since Darwin.

I propose to augment accepted evolutionary theory, the theory that Darwin (1859) summed up in the phrase "descent with modification." Modern neo-Darwinists might prefer to express it as "descent with genetic mutation and recombination," but they would agree with Darwin (p. 455) that "it does not seem incredible that . . . all the organic beings which have ever lived on this earth may be descended from some one primordial form." He also made it clear that, in his opinion, the line of descent was always from variety to variety, the accumulation of varieties leading to new species, new genera, and, ultimately, new phyla and new kingdoms (pp. 160–161). Of course Darwin also put forward the concept of natural selection to explain why organisms with adaptive modifications survived whereas others, less well adapted, became extinct. The conventional view is that all organic evolution is the result of the selection of slight, adaptive, inherited modifications, and Darwin stressed the importance of slight modifications. "Why should not Nature have taken a leap from structure to structure?" he wrote (1859, p. 223).

"On the theory of natural selection, we can clearly understand why she should not; for natural selection can act only by taking advantage of slight successive variations; she can never take a leap, but must advance by the shortest and slowest steps." The views that I am now putting forward suggest that nature has taken occasional leaps, not from structure to structure but from life history to life history. These leaps do not involve leaps in genetic mutation or macromutations, but they do involve leaps in genetic constitution by combining complete genomes. As far as is known, this is a concept Darwin never contemplated. I now ask biologists to consider the possibility that Darwinian evolution may not be the only process in animal phylogeny, or processes other than descent from variety to variety or from species to species may have played a significant part in determining the genealogy and relationships of many animals.

I am certainly not the first to suggest that the accumulation of mutations is not the only process to have played a part in the phylogeny of organisms. There is now growing acceptance of the hypothesis that "several prokaryotes make a eukaryote" (Margulis, 1981), or the ancestry of all cells, including those of plants and animals, can be traced back to symbiotic associations between different types of bacteria (prokaryotes). Although this concept originated quite early in the present century, it was largely ignored until Lynn Margulis accumulated evidence from many sources and critically assessed it (Margulis, 1970, 1981). Symbiosis, however, does not seem to provide an explanation of the occurrence of different types of larvae in the animal kingdom, and I suggest that there is another process that has brought together distantly related evolutionary lineages and combined them to produce species with new life histories.

My new hypothesis of transfer of larval form was developed solely from morphological and developmental studies. Recent developments in a quite different branch of biology, however, may be regarded as enhancing the credibility of my views. I am referring to genetic engineering and gene sequencing, both of which are aspects of molecular biology.

Genetic engineering covers a number of laboratory techniques whereby one or a few genes from one organism are transferred to cells of a second organism, which need not be closely related to the first. Thus the gene that produces insulin in human cells can be transferred to a bacterium or a mouse egg, and the offspring of the bacterium, or the mouse that grows from the egg, will then produce human insulin (Cherfas, 1982). This form of genetic engineering is capable of transferring only small packages of genetic material between unrelated cells, and I doubt that a comparable natural process could have been responsible for the transfers of the large amounts of genetic material that I postulate. It does show, however, that if genes can be transferred to cells of very distantly related species they are not necessarily rejected by those cells and they can function to produce their original product. Larger transfers of genetic material have been achieved by somatic cell hybridization, which can take place under some conditions when distantly related cells are grown in culture (Rothwell, 1983). The necessary conditions probably never occur in nature, so again the process is unlikely to be involved in the genetic transfers that I postulate. Its relevance, however, is that it demonstrates that large quantities of genetic material from widely different sources can be incorporated in viable cells.

Gene sequencing involves the analysis of stretches of genetic material into their chemical building blocks, the nucleotides. In some instances remarkably similar sequences have turned up in distantly related species, and this has led some workers in the field to consider the possibility that gene clusters have been transferred (Lewin, 1982; Syvanen, 1985; Marshall, 1988). We do not know what features these genes are coded for, and we can only speculate as to how they might have been transferred, but it does seem that independent evidence from parts of molecules within cell nuclei and from whole animals in their different stages of development is consistent with the concept that relatively large transfers of genetic material have taken place between species that are not closely related.

The transfer of genetic material between distinct evolutionary lines has been termed "horizontal genetic transfer," alluding to horizontal links between branches of the phylogenetic tree. The horizontal transfer of larval form that I now postulate is ascribed to horizontal transfer of the whole genome of one animal to another. Horizontal genome transfer is considered primarily as a process leading to new life histories, but with great consequences for survival, dispersal, and speciation. It may be much rarer than descent from species to species, but I suggest that in several major groups of animals its evolutionary impact has been no less.

People had been classifying plants and animals into species, genera, families, and larger groups in a so-called "natural system" long before there was any general acceptance of the concept of evolution. It was widely recognised, for example, that whales and porpoises were mammals, even though they swam like fish, and bats were also mammals, even though they flew like birds. Darwin (1859) was able to show that this natural system of classification was the result of evolution, and that "propinquity of descent,—the only known cause of the similarity of organic beings,—is the bond, hidden as it is by various degrees of modification, which is partially revealed by our classifications" (p. 399). He showed that descent with modification and relatedness could be summarised in a form of branching diagram. His own example (pp. 161–162) was based on hypothetical varieties, species, and genera, but he left no doubt that it was at least theoretically possible to draw one or more "trees of life" covering the genealogies and relationships of all life, living and extinct. His followers soon produced them. A standard example of a tree generally acceptable to zoologists is shown in Figure 3.1.

Most modern biologists believe that the evolutionary relationships of organisms can be shown in some form of dendrogram, which is merely a general term for a branching, tree-like diagram. There are disputes about how to obtain a diagram that correctly depicts the true relationships of organisms, and further disputes as to whether and to what ex-

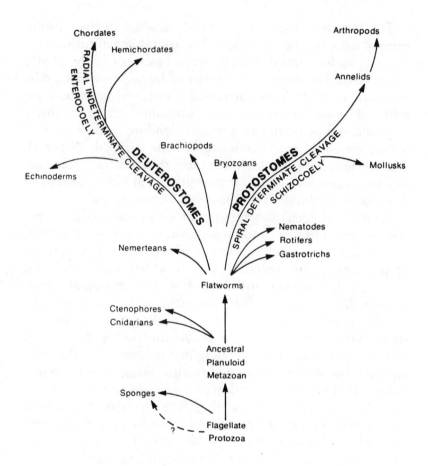

FIG. 3.1. A phylogenetic tree of the animal kingdom. (From Barnes, 1980.)

tent classifications should be derived from dendrograms. Most modern methods have sought to reduce the subjective or intuitive element in arriving at either a classification or a phylogeny, but they have approached the problem in very different ways and arrived at very different results. This is illustrated by both the numerical taxonomists and the cladists. Numerical taxonomists take the view that the construction of a classification should be quite independent of any deductions on phylogeny. In arriving at a classification they originally advocated treating all characters as of equal importance, using vast numbers of them and getting a computer to sort out degrees of similarity. As critics of the system soon pointed out, one could never use all possible characters of organisms, so there is inevitably some selection, either conscious or unconscious; giving all characters equal weight is itself a form of weighting; and is it really defensible to regard the colour or size of an organism as of equal importance to its cellular structure or lack of it? Cladism itself has evolved into several groups, but all insist that the phylogenetic tree, which they call a cladogram, must be constructed according to certain rigid rules. Most consider that all classifications should be based strictly on cladograms, and that any classificatory group must include all the descendents of a common ancestor. Critics of this method suggest that some of the rules of cladistic analysis are themselves derived subjectively and question whether the resulting cladograms always reflect true phylogenies. They also point out that by deliberately ignoring degrees of difference between organisms the cladists are reducing the usefulness of the resulting classifications. Today both numerical taxonomists and cladists have modified their original positions, but there is no general consensus among biologists as to the principles that should be employed in classifying organisms or in deducing phylogenies. Few, however, would dispute that similarities between organisms are usually the result of inheritance and differences usually the result of evolution.

I am well aware that in the preceding paragraph I have not done full justice to anyone's views on classification or phylogeny, whichever faction he or she belongs to or opposes. That is not my object. I do wish, however, to point out that all methods of classifying organisms depend on the evolutionary relationships of the organisms, for, even in numerical classifications, the number of shared characters is itself a measure of relatedness. Darwin and most later biologists have regarded the clues from embryos and larvae as among the most important in deducing the relationships of animals, but I am now suggesting that these clues can often be misinterpreted and that the accepted views on the relationships between the major groups of animals are, in many cases, wrong. The relatedness of animals, I claim, can be rather more complex than has previously been suspected. Much of it does depend on the evolution of species from other species, as cladograms and other dendrograms imply, but I maintain that it can be complicated by the occasional transfer of genes between distantly related groups and that these genetic transfers particularly affect the form of the embryonic and larval stages. This implies that the branches of a dendrogram can occasionally interact, a situation difficult to show on conventional diagrams. I regard these genetic transfers between distantly related animals as being occasional, in that they have produced significant results at rather infrequent intervals. In many cases, however, these results have been very significant indeed. They have affected more than half the phyla of the animal kingdom, and the process is continuing. My point, for the moment, is that this is not a matter that concerns only students of embryos and larvae; it is also of the utmost importance to all zoologists who use animal classifications.

Darwin was rightly insistent that species consist not only of adults but of all stages in development, for all stages must survive until reproduction has taken place for the organism to perpetuate itself, and all stages evolve or are capable of evolution. Darwin also insisted that, for animals, the characters of embryos and larvae are every bit as important as those of adults as indicators of relationships. Now Darwin assumed, and until a few years ago I would have agreed, that any animal has only one line of descent, from species to species, and that it is legitimate and sensible to try to deduce its extinct ancestors and living relatives from evidence from all possible sources. Darwin's own knowledge of embryos and larvae was fully consistent with the idea that pieces of phylogenetic evidence from all phases of development must be fully compatible with each other. He cited the case of the cirripedes or barnacles, a group on which he had published an authoritative monograph some 8 years before *The Origin of Species*. The best known barnacles live in calcareous shells, cemented to rocks, and in Darwin's youth they had generally been regarded as molluscs. "Even the illustrious Cuvier" he wrote in the *Origin* (p. 420) "did not perceive that a barnacle was, as it certainly is, a crustacean; but a glance at the larva shows this to be the case in an unmistakable manner." In fact, a barnacle hatches from the egg as a nauplius (Fig. 3.2a, b), a type of larva that occurs very widely among crustaceans but that is unknown in any other group. Then if we look again at the adult, inside the calcareous shell we find an animal enclosed in a chitinous cuticle, with jointed limbs not unlike those of a shrimp, and, like a shrimp, it moults this cuticle at intervals and the body expands a little before the next cuticle hardens. So the evidence from the larva and the adult is fully compatible: a barnacle is a crustacean, not a mollusc.

During the nineteenth century there was an increasing interest in the life histories of marine animals, but by the time *The Origin of Species* was published the number of accurate descriptions of reliably identified marine larvae was still quite small. Thus, still on the subject of his own speciality, Darwin stated (1859, p. 420) that "the two main divisions of cirripedes, the pedunculated [stalked] and sessile, which differ widely in external appearance, have larvae in all their several stages barely distinguishable." In fact the nauplius larvae of stalked barnacles have more and longer processes than do the corresponding stages of sessile species (Fig. 3.2a, b), so that the larvae of the two groups are readily distinguishable. The two types of nauplii both show the main features of barnacle larvae, so there is no conflict between the affinities of adults and larvae, but the example is sufficient to show that Darwin's knowledge of marine larvae was limited. The assumption that the affinities of different phases in development would always be compatible was fully consistent with what Darwin knew, but this is a questionable assumption in light of present knowledge.

As more and more larvae have become known, biologists have sought to reconcile the affinities of the larvae with those of the corresponding adults. In many cases classifications suggested by the larvae and by the adults have been entirely

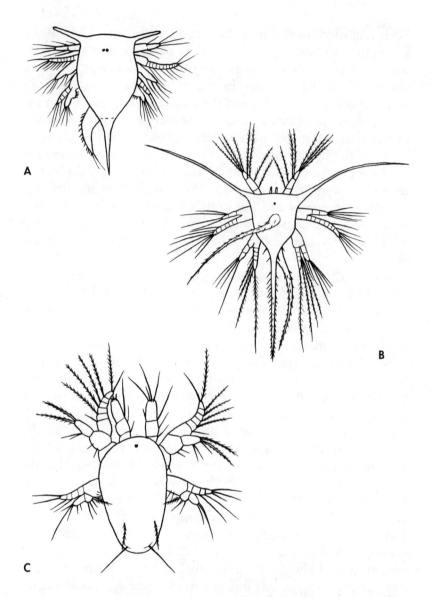

A

B

C

FIG. 3.2. Some crustacean larvae. (a) Nauplius of *Balanus,* an acorn barnacle (Cirripedia); (b) nauplius of *Lepas,* a stalked barnacle (Cirripedia); (c) nauplius of *Cyclops* (Copepoda); (d) zoea of *Pandalus* (Caridea); (e) zoea of *Pagurus* (Anomura); (f) zoea of *Cancer* (Brachyura). (a–c) Dorsal views; (d–f) lateral views with dorsal views of telsons. (a, b redrawn from Tregouboff and Rose, 1957; c redrawn from Gurney, 1942; others original.)

LARVAE AND EVOLUTION

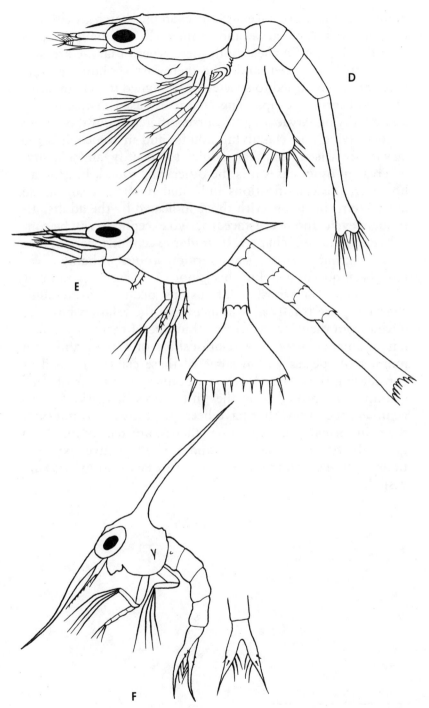

D

E

F

in accord. Most crustaceans, for example, hatch as either a nauplius or a zoea larva, and some hatch as a nauplius and later develop into a zoea. Both these types of larvae are limited to the Crustacea, and, like their corresponding adults, they can be identified to species and assigned to genera, families, and larger groups. The characters used in identifying and classifying larvae are, of course, different from those used in identifying and classifying adults, and in some cases the larvae are easier than the adults to identify and classify, whereas in other cases it is the other way round. In general, however, the classifications indicated by crustacean larvae are quite reconcilable with those indicated by the adults, although, even among crustaceans, two exceptions will be considered in the next chapter. If similar exceptions were as rare in other groups Darwin's assumption would probably never have been questioned, but this is not so. There are a number of whole phyla whose relationships deduced from adult characters are totally at variance with the relationships deduced from their larvae, and within one of these phyla similar anomalies affect the relationships of classes and of a number of species within some of these classes. Instead of considering these paradoxes in isolation, a number of such examples are brought together in the present work. The ingenious suggestions that have been put forward to interpret them in terms of conventional theory are reexamined and generally found inadequate, and my alternative explanations, in terms of horizontal transfer of larval form, are suggested.

II
EXAMPLES

4
Decapod Crustaceans

Differences between Brachyura (true crabs) and Anomura (including hermit crabs)—Adult Dromiodea are brachyurans but their zoea larvae are anomurans and resemble larvae of hermit crabs—Explanations involving misclassification and larval conservatism rejected—Suggestion that early dromioid, with no zoeal phase, acquired larvae from early hermit crab by hybridization—Adult *Dorhynchus* is a spider crab of family Inachidae but zoea larvae share characters of Inachidae and Homolidae—Comparable anomalies occur in some echinoderms and all are explicable as result of hybridization

The main subject of this book is the postulated transfer of embryonic and larval forms from one phylogenetic line to another, and I start with the decapod crustaceans. This is a group that contains very few examples that can be attributed to this phenomenon, but the case of the Dromioidea is particularly important. It presents the paradox of brachyuran adults and anomuran larvae, so well known among carcinologists but, until now, totally unexplained. There is a possible alternative explanation for my other example in this group, the anomalous larva of *Dorhynchus*, so had I restricted my attention to decapod crustaceans I would have had only one example that seemed to demand a new hypothesis. When, however, I eventually came to consider groups in other phyla I found no shortage of paradoxes that could be explained as examples of the transfer of embryonic and larval forms from distinct lineages. For some of the paradoxes no other explanation has been offered, and for others I regard the accepted explanations as weak or flawed.

In the two examples known to me of crustacean larvae that seem inconsistent with their corresponding adults, the adults are both crab-like. The Crustacea were usually regarded as a class of animals until about 20 years ago (e.g.,

Borradaile et al., 1935), but some more recent texts classify them as a subphylum (e.g., Barnes, 1980) or even a phylum (e.g., Barnes et al., 1988). The group covers a great variety of forms, but the only ones that concern us here are the Brachyura and the Anomura, both of which are groups of the order Decapoda. All the more familiar shrimps, lobsters, and crabs are classified in this order. The name refers to their 10 (five pairs) of legs, appendages that are typically used for walking, although some or all of them may be modified for other purposes. The name Brachyura means "short tails," and this group comprises the true crabs, in which the abdomen (tail) is always quite small and fits into a groove under the thorax. The head and thorax are fused together and are often very broad. The female always has a receptacle for storing spermatophores received from the male so that mating may preceed egg-laying. Females of some families of crabs mate only once, but egg-laying may take place at intervals over several years. The Anomura are the "intermediate tails," in which the abdomen is usually much less reduced and more readily extensible than in the Brachyura. Female anomurans lack means of storing sperm, so the eggs can be fertilized only if a male is present when they are laid. Within this group, the hermit crabs (families Paguridae and Diogenidae) are particularly relevant in the present context. The abdomen is quite large and without a hard shell, but the hermit crab usually protects it with the empty shell of a gastropod mollusc.

Most brachyurans and anomurans hatch as zoea larvae, that is larvae that swim with the outer branches (exopods) of some of the thoracic appendages, but the larvae of the two groups are quite distinct. The carapace of a typical brachyuran zoea (Fig. 3.2f) is almost spherical and usually bears a ventrally pointing rostrum, a dorsal spine, and a lateral spine on either side, although in some cases one or more of these spines is missing. The telson is in the form of a two-pronged fork. The carapace of a typical anomuran zoea (Fig. 3.2e) is more elongated, and the rostrum points forward; dorsal and lateral carapace spines are very rare but a posterior spine on either side is very common. The telson is a flat plate, often triangular, on which the outermost process is a short spine and the one next to it a fine hair.

For over a century carcinologists have been trying with little success to reconcile the adult and larval features of *Dromia*, the sponge crab (Fig. 4.3a,b) (see Williamson, 1988b). The name "sponge crab" refers to the habit of carrying a sponge over the back, holding it with the last two pairs of legs. This and related genera, which together make up the superfamily Dromioidea, look like early evolved members of the Brachyura. Their zoea larvae, on the other hand, bear little resemblance to those of the Brachyura (compare Figs. 4.3b and 3.2f) but show a very close resemblance to larvae of the Anomura (Fig. 3.2e), particularly to the larvae of hermit crabs (Paguridae and Diogenidae). Now there are a number of anomurans, including the stone crabs (Lithodidae) and the porcelain crabs (Porcellanidae), which, as adults, superficially resemble true crabs. They have a broad carapace and a relatively small abdomen that bends forward under the thorax, just as in true crabs, but there any resemblance to true crabs ends. They show clear resemblances to the Anomura rather than the Brachyura in the form of the larva, in the arrangement of the endophragmal skeleton (which forms attachments for muscles and supports some internal organs), and also in the fact that the female has no spermathecae to store spermatophores from the male. There is no argument over the Lithodidae or the Porcellanidae: they are universally accepted among biologists as anomurans, which, as adults, look superficially like crabs. They do show, however, that it is possible to look superficially like a true crab and yet not be one.

Are the dromioids true brachyurans, or are they another example of anomurans that superficially resemble brachyurans, like the lithodids and porcellanids? This is the question that has been puzzling carcinologists for so long, and it puzzled me for over 30 years. For most of that time my view was expressed by adapting Darwin's comment on barnacles, quoted in the previous chapter. In my version, I would say

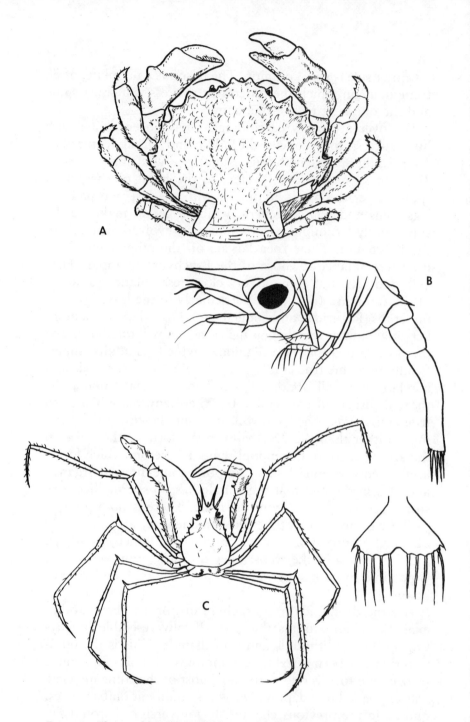

FIG. 4.1. Three crabs with unusual larvae. Adults in dorsal view, larvae in lateral view; to different magnifications. (a, b) Adult *Dromia* and stage I zoea, with dorsal view of zoeal telson; (c, d) adult *Dorhynchus* and last zoeal stage; (e, f) adult *Homola* and last zoeal stage. (a redrawn from Monod, 1956; b redrawn from Gurney, 1942; c redrawn from Thomson, 1874; d original; e redrawn from Rice and Provenzano, 1970; f redrawn from Rice and von Levetzow, 1967.)

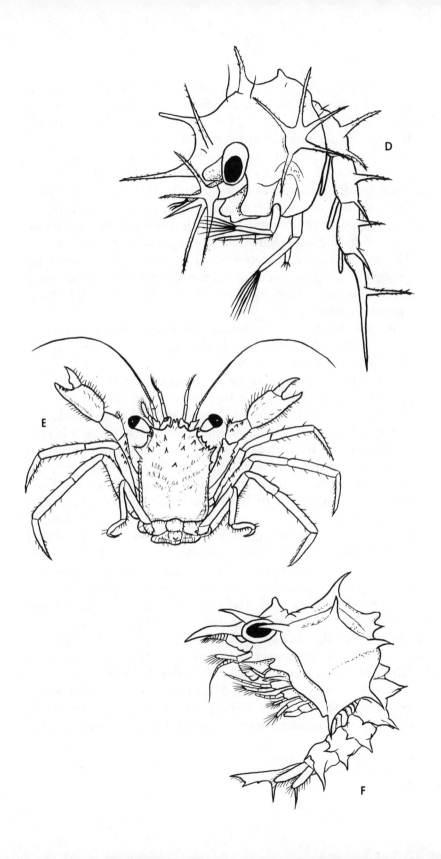

D

E

F

"even the illustrious Calman did not perceive that a drom-ioid was, as it certainly is, an anomuran, but a glance at the larva shows this to be the case in an unmistakable manner." It must be admitted, however, that not a single student of adult dromioids has been swayed by this assertion, and I am now convinced that the cases of the barnacles and the drom-ioids are not parallel. Detailed examination of an adult bar-nacle shows it to be a crustacean, just as its larvae suggest, but detailed examination of an adult dromioid does not show it to be an anomuran, in spite of the form of its zoea larvae. The endophragmal skeleton of adult dromioids is essentially brachyuran, not anomuran, and the females of the group have spermathecae resembling those of early brachyurans but un-known in anomurans.

Recent descriptions of adult and larval dromioids merely confirm the situation: the adults really do belong to the Brachyura and the zoea larvae to the Anomura. There can be from one to six zoeal stages in different species, and the last zoeal stage will moult to produce a megalopa, a larva that swims with its abdominal appendages before eventually settling. The megalopa is the final larval phase, and it is suc-ceeded by the first juvenile stage. Not all dromioids, how-ever, pass through a zoeal phase in their development, for many hatch as a megalopa. The known megalopas of the Dromioidea all seem to have brachyuran rather than ano-muran characters, but it must be admitted that the distinc-tions between the two groups are much less clearly defined in megalopas than in zoeas, juveniles or adults. One sugges-tion is that the Brachyura evolved from anomuran ancestors and the Dromioidea somehow managed to keep the ancestral form of zoea larva while the adults evolved brachyuran fea-tures, but I reject this explanation. Apart from dromioid lar-vae, all the other evidence, from both adults and larvae, sug-gests that the Brachyura did not evolve from an anomuran ancestor, but the evolutionary lines that led to the Anomura and the Brachyura diverged separately from lobster-like ancestors (Guinot, 1978; Williamson, 1988b). The characters of adult Dromioidea, both extant and fossil, are quite consistent with evolution from this lobster-like ancestor. Most dromioid zoea larvae, however, show a mixture of the zoeal characters of the two families of hermit crabs, the Paguridae and the Diogenidae. In the only known exceptions, some features of dromioid zoeas are not those of present-day hermit crabs but features that probably occurred in their ancestors. Species of *Dromia*, for example, develop more thoracic exopods than the larvae of hermit crabs, but there has been a general evolutionary tendency in many groups of decapod crustaceans to reduce the number of thoracic exopods (Williamson, 1988b). The larval evidence is therefore consistent with the evolution of the Dromioidea from an ancestor that also gave rise to the Paguridae and the Diogenidae. This common ancestor was probably an early hermit crab, but it was much more evolved than the common ancestor of the Anomura and Brachyura.

If the most recent common ancestor of both the Anomura and the Brachyura had a larva of the dromioid type, we should expect other brachyurans that have retained a large number of ancestral features to have similar larvae, but they do not. This applies particularly to the Homoloidea, in which some adult characters, including the endophragmal skeleton, are less evolved than those of the Dromioidea (Guinot, 1978). This group contains Jurassic representatives, and is regarded by Guinot and others as containing the ancestors of the more evolved Brachyura. This view is supported by the evidence from the larvae, for intermediate steps are known connecting the characters of homoloid larvae (Fig. 4.3f) with those of more typical brachyuran larvae (e.g., Fig. 3.2f) (Williamson, 1976, 1988b). The known facts seemed to suggest that an ancestor of all the modern dromioids might have "acquired" the basic form of its zoea larva from an anomuran, but how could one larval form supplant another and leave the adult form unaffected?

Most students of fossil and living adult crabs believe that the early Dromioidea either included the ancestors of the modern Brachyura or were closely related to them. In either case we should expect their larvae to resemble those from which modern crab larvae evolved, and I suggest that early dromioids had such larvae. It has already been pointed out, however, that among the modern Dromioidea there are many species that hatch from the egg as a megalopa, and a zoeal phase, which normally precedes the megalopa, does not occur in the life history. I suggest that this tendency to suppress the zoeal phase developed quite early in the evolution of the group and that an early female dromioid with no zoeal phase (or rather the eggs of such a female) acquired the genetic prescription for zoea larvae from an early hermit crab. All modern dromioids with anomuran-like zoea larvae are then explicable as descendents of this female. Fossil forms quite similar to some modern dromioids are known from Jurassic strata, and the transfer of genes coded for the new larval phase might have taken place during that geological period or in any subsequent period. According to my hypothesis, then, dromioid larvae might have had anything up to 150 million years to reach their present state of diversity. I suggest that cross-fertilization presents the most probable method of acquiring the genetic material to produce a new larval form without necessarily affecting the later phases in the life history, and inserting a new larval phase at the beginning of a life history seems to present a simpler genetic problem than substituting one larval form for another. The probable method is, however, quite distinct from the suggestion that new larval forms have been introduced into the life histories of many groups. For the moment I prefer to concentrate on the evidence that larval transfers have taken place, and later (Chapter 12) I shall discuss the possible mechanisms.

Another crustacean is known whose larvae show some very unusual features that also occur in the larvae of a different superfamily, but its other larval features are much as one might expect from the adult. This is *Dorhynchus thomsoni* (Fig. 4.3c, d), a small deep-sea spider crab, whose larvae I first described in 1960 but could not identify with any certitude until 1982. The adult *Dorhynchus* seems quite clearly to be a member of the family Inachidae (superfamily Majoidea), but although some of the characters of the larvae agree with this family others are strikingly different. Zoea larvae of brachyuran crabs typically have four carapace spines (including the rostrum), but one consistent feature of inachid zoeas (apart from *Dorhynchus*) is that they have only a dorsal carapace spine and no laterals or rostrum. *Dorhynchus*, by contrast, has 14 carapace spines and two large tubercles, and the abdomen is also unusual in having two pairs of spines on most segments instead of one. The telson and appendages, however, are very similar to those of other inachid zoeas, with the same distribution of spines and setae, although the spines on the antenna and telson are unusually long. (Most of the features of the telson and appendages are not shown in Fig. 4.3 but are described and figured in Williamson, 1982.) The 14 carapace spines include not only a dorsal spine, which might be expected, but also a forwardly directed rostrum, a spine over each eye, and a rosette of five spines on each side, all of which are quite unexpected in this family. Indeed, this combination of carapace and abdominal spines is known elsewhere only in the late zoeal stages of *Homola* (family Homolidae, superfamily Homoloidea) (Fig. 4.3f). The zoea of *Homola* has already been mentioned as being very different from the larvae of dromioids and hermit crabs. As far as its telson and appendages are concerned, it is also quite different from the larva of *Dorhynchus*, but the fact that each carapace and abdominal spine of the larva of *Dorhynchus* has its counterpart in the larva of *Homola* is much too striking to be dismissed as mere chance.

The Homoloidea contain fossil forms from which most present-day crabs, including the Inachidae, probably evolved (Guinot, 1978). As modern homoloids seem to have retained many of the features of this ancestral stock, it also seems quite likely that their larvae should have done so. If modern inachid zoeas evolved from forms like the zoea of *Homola*, it seems just possible that they have retained the genetic recipe for homoloid carapace and abdominal spines, although the relevant genes are suppressed in the great majority. If this were the case, the carapace and abdominal spines of crab zoeas would provide a parallel case to bird's teeth, for, although no modern birds have teeth, they apparently have retained the genetic prescription for them from their reptilian ancestors, and teeth can sometimes be obtained in cultures of avian tissue. (Bird's teeth and other examples of atavism are reviewed by Hall, 1984.) Perhaps something has happened to *Dorhynchus* that has reactivated the suppressed genes for additional carapace and abdominal spines. I previously favoured this explanation (Williamson, 1988a), but I have now realised that there are quite similar cases in another phylum that cannot be explained in this way. Some species of sea-urchins, discussed in Chapter 8, share some but not all the characters of distantly related species within the same class of animals, although in these examples the shared characters are probably not ancestral and the reactivation theory does not seem applicable. I now suggest that horizontal genetic transfer will, in cases like the Dromioidea, result in the introduction of a new larval form, whereas in other cases, like *Dorhynchus*, it will introduce some new characters to an existing larva. The details of this mechanism are discussed in Chapters 12 and 13.

5
Echinoderms: Adults and Larvae

Adult echinoderms described—Adult echinoderms radially symmetrical, larvae bilaterally symmetrical—Conventional explanation assumes ancestral echinoderms bilaterally symmetrical throughout life but adults secondarily adopted radial symmetry—This view criticised on theoretical grounds and for lack of evidence—Partial bilateral symmetry of holothurians not inherited from ancestors of all echinoderms but probably a recent case of neoteny—Alternative view: original echinoderms radially symmetrical throughout life, without planktonic larvae; bilateral larvae transferred from another phylum; further transfers within echinoderms

The Echinodermata are a major phylum of marine animals, and they provide many striking examples of inconsistencies between the apparent relationships of adults and larvae. These inconsistencies occur in trying to assess the relatedness, first, of the echinoderms to other phyla, second, of classes within the echinoderms, and, finally, of several pairs of species within the same class. I hope to show that these inconsistencies are resolved by the suggestion that some larval forms were acquired from animals in distinct evolutionary lineages. Before this, however, let us consider what the animals are like, both as adults and larvae.

All living adult echinoderms are radially symmetrical. They have an oral side (with a mouth) and an aboral side (opposite the mouth), but there is no dorsal and ventral, left and right. Jellyfish and their relatives provide other examples of radial symmetry among animals, but many more examples, including a great many very familiar ones, occur among plants. Thus a daisy is a radially symmetrical flower and an orange is a radially symmetrical fruit. Now whereas an orange is composed of a large and variable number of similar radial sectors, a typical echinoderm is composed of only five, so it is pentaradial. There are sometimes more than five radial elements, but they usually occur in multiples of five. The name echinoderm means spiny skin, and nearly all echinoderms do have calcareous spines. Some parts of the outer surface also bear soft, flexible podia, the tube-feet, which are connected to an internal system of fluid-filled canals, the water vascular system. This system is distinct from the main coelom, through which the gut runs and into which the gonads protrude. The coelomic fluid is circulated by cilia, and there is also a haemal system of channels with a rather simple, contractile heart. Members of only one class of echinoderms have a specialised respiratory pigment. Nerves radiate from a circumoral ring and branch to all parts in contact with the outer environment, but there is no concentration of nerve cells that can be regarded as a brain. There are no special organs of excretion. Eggs and sperm are usually shed into the sea, where fertilisation takes place, but in a minority of cases eggs are fertilised in the coelom.

Living echinoderms are divided into six classes, which I call the Asteromorpha, Ophiuromorpha, Echinomorpha, Holothuromorpha, Crinomorpha, and Concenticyclomorpha (Williamson, 1988a). The more traditional names for all these classes end in -oidea rather than in -omorpha, but the ending -oidea is also used throughout the animal kingdom for the names of superfamilies, and this usage has the backing of the International Code of Zoological Nomenclature (ICZN, 1985). It is an unnecessary confusion to have identical names for classes and superfamilies, and a change seemed overdue.

The Asteromorpha (Fig. 5.1a) are the starfish or sea stars, with typical genera such as *Asterias*, but also including sun stars such as *Solaster* and cushion stars such as *Asterina*. They usually have five arms, but they can have up to 50. The mouth is in the middle of a central disc and faces downward, and there is usually an anus on the opposite side of the disc. The arms taper outward from the disc, and along the oral surface they have rows of tube-feet with suckers, which allow the animal to creep over the surface. The arms are relatively flexible but move slowly. Branches of the gut, the gonads, the main coelom, and the water vascular system extend into each arm. Finger-like extensions of the coelom, termed papulae, are restricted to this class. They protrude slightly from the aboral surface and act as little gills. The pedicellariae are small pincer-like appendages, scattered over much of the surface, which capture small prey and remove debris. Such appendages are also found in the Echinomorpha. Most starfish are predators or scavengers, and the food of different forms includes living and dead molluscs, crustaceans, fish, and other starfish. One large species, known as "the crown of thorns," feeds on corals and has done extensive damage to parts of the Great Barrier Reef of Australia.

The Ophiuromorpha (Fig. 5.1b) are the brittle-stars, with such genera as *Ophiura* and *Ophiothrix*. Five arms are usual, as in starfish, but the arms are marked off much more clearly from the central disc than those of starfish, and they are much more slender and more flexible. The name "brittle-star" refers to the tendency of the arms to break when the animal is trapped or handled. The prefix "ophi-" or "ophio-," which occurs in the name of the class and so many of its genera, refers to the snake-like form of the arms. The arms are protected externally by spines and skeletal plates, and they also have an internal skeleton of jointed ossicles, sometimes termed vertebrae. There are longitudinal muscles and nerves in the arms, but only a very restricted coelomic cavity and no branches of the digestive or reproductive organs. Brittle-stars move by pulling themselves around with their arms, rather than by creeping on their simple tube-feet, which have no suckers, and some can swim a little by waving the arms. The mouth, on the under surface of the disc, leads to a simple stomach, which ends blindly: there is no intestine and no anus. Some feed on bottom detritus, and others, which live in tidal currents, fish by holding up one or more arms to catch live plankton or dead suspended matter.

The Echinomorpha (Fig. 5.1c) are the sea-urchins, such as

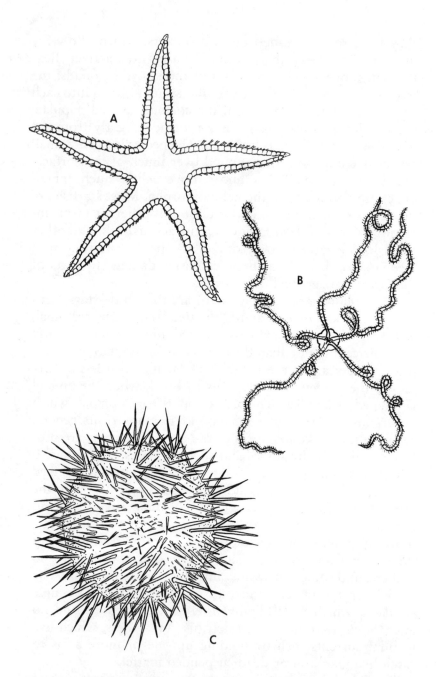

FIG. 5.1. Examples of adults of the six living classes of echinoderms. (a) *Astropecten* (Asteromorpha); (b) *Acrocnida* (Ophiuromorpha); (c) *Echinus* (Echinomorpha); (d) *Cucumaria* (Holothuromorpha); (e) *Antedon* (Crinomorpha); (f) *Xyloplax* (Concentricyclomorpha). (a–c redrawn from Bell, 1892; d, e redrawn from Borradaile et al., 1935; f redrawn from Baker et al., 1986.)

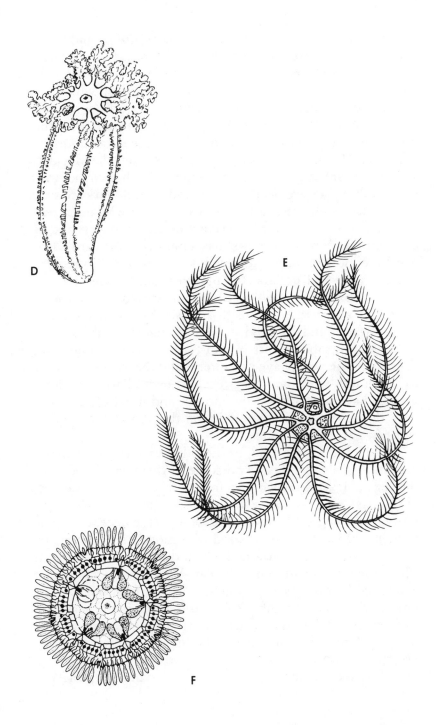

D

E

F

Echinus, heart urchins, such as *Echinocardium,* and sand dollars, such as *Mellita.* They have no arms, and the body is enclosed in a rigid shell of closely fitting plates that support movable spines and pedicellariae and through which the tube-feet protrude. Regular sea-urchins are almost spherical, with the mouth in the middle of the surface that faces the substratum; the anus, surrounded by five genital pores, is opposite. The gonads of some species are considered a delicacy in several countries. Heart urchins take their name from the fact that they are heart shaped, and sand dollars are flattened discs. Both these latter groups have some bilateral features superimposed on their original radial symmetry, and in the heart urchins the mouth has migrated "forward" (toward the indentation) and the anus "backward." In regular urchins and sand dollars, the mouth is surrounded by five teeth flexibly supported by a skeletal structure known as "Aristotle's lantern." The teeth are used by urchins to graze hard surfaces and by sand dollars to ingest small particles found in the sand. Heart urchins, which live in tubes that they dig in the sand, have no lantern, and they move particles to the mouth with their tube-feet and pedicellariae. Regular urchins have gills around the mouth, and heart urchins and sand dollars have leaf-like podia on the aboral surface that assist in gaseous exchange.

The Holothuromorpha or holothurians (Fig. 5.1d) are known as sea cucumbers, and some of them are shaped like the vegetable from which they take their common name and, in the case of *Cucumaria,* their Latin name also. The similarity to the vegetable does not extend to the taste, but some of them are used as human food. The mouth, surrounded by tentacles, is at one end of the body and the anus at the other. The body wall is soft and muscular. Very small denticles are embedded in this cuticle, but there are no surface spines. Some holothurians have no tube-feet on the trunk, but others have them scattered over the surface or arranged in five rows. The tentacles round the mouth are modified tube-feet and vary greatly in size and form in different orders. All members of this class show some degree of bilateral symmetry, and this is seen most clearly in creeping forms, which always move with the same side of the body in contact with the substratum. There is only one gonad in all holothurians, and it occurs on the upper side of creeping forms. The single gonad is the only obvious departure from radial symmetry in burrowing forms, such as *Synapta,* and also in the only pelagic form, *Pelagothuria,* which swims with long tentacles, webbed at the base.

The Crinomorpha (Fig. 5.1e) contain the only living fixed echinoderms, the stalked sea lilies, such as *Cenocrinus.* The class also contains the free-living feather stars, such as *Antedon,* although *Antedon* also has a stalk when young. The body is enclosed in calcareous plates. The stalk, when present, arises from the aboral side, and even when there is no stalk the mouth normally points upward. The anus is close to the mouth. Slender, flexible arms fan out from the body in a basically pentaradial formation, but they are usually much branched and have a feathery appearance. The arms are used to collect living or dead plankton, which is conveyed to the mouth by ciliary currents, but in free-living forms they also enable the animal to crawl over the bottom or even to swim with gentle waving movements.

The class Concentricyclomorpha (Fig. 5.1f) was set up in 1986 (as the Concentricycloidea) to accommodate a new species, *Xyloplax medusiformis,* discovered in sunken, waterlogged wood in deep water off the coast of New Zealand. *Xyloplax* has been called a sea daisy because of its disc-shaped body surrounded by petaloid spines. It has a ring of tube-feet, served by a double ring of water vascular canals, there is no mouth or gut, and the embryos develop directly to forms like miniature adults, with no larval stage.

Larvae are, by definition, very different from adult animals, but the echinoderms provide the only known examples of bilaterally symmetrical larvae that give rise to radially symmetrical adults. Not all echinoderms have larvae, but the majority of them do. In those with larvae, the length of larval life is a matter of days in some species, and weeks or months in others. Typical larvae, such as those that occur in plank-

ECHINODERMS: ADULTS AND LARVAE

ton samples in practically all seas and oceans of the world, have not only an oral and an aboral side, as do the adults, but also a left and a right and an anterior and a posterior, which the adults do not. Even the oral and aboral sides of the larva do not correspond to the oral and aboral sides of the adult, but the method of changing from the larva to the adult is such a peculiar performance that it deserves to be treated separately.

The egg of any metazoan animal is radially symmetrical and usually spherical, whatever shape of animal it comes from. Fertilization initiates the process of cell division, which soon leads to the formation of a hollow ball of cells, the blastula (Fig. 5.2a). In a great many echinoderms, the blastula is ciliated, and the egg hatches at this stage. In some, however, hatching is delayed until further development has taken place. The developing animal, hatched or unhatched, now changes from a blastula to a gastrula (Fig. 5.2b). The early echinoderm gastrula is still radially symmetrical, but some bilateral features develop during the late gastrula stage. Opposite sides of the inner end of the archenteron become produced into pouches (Fig. 5.2c), and these seal off to form pairs of fluid-filled vesicles (Fig. 5.2d). Up to three pairs of sacs are formed in this way and provide an enterocoel as the blastocoel fills with cells. The coelomic pouches and resultant sacs are lateral to the archenteron, but it is difficult to say which side is right and which is left until the mouth begins to develop on the ventral surface. The mouth forms a connection with the inner end of the archenteron; the blastopore becomes the anus (Fig. 5.2e). An echinoderm larva is thus a deuterostome, because its mouth is not derived from the blastopore. Deuterostomy is usually associated with enterocoely.

At about the same time as the developing echinoderm larva acquires a mouth, the surface cilia become concentrated in a band around the mouth (Fig. 5.2e). As development proceeds this band is greatly extended, and its form varies considerably among the different groups of echinoderms. It may become much convoluted or it may give rise to several distinct bands.

LARVAE AND EVOLUTION

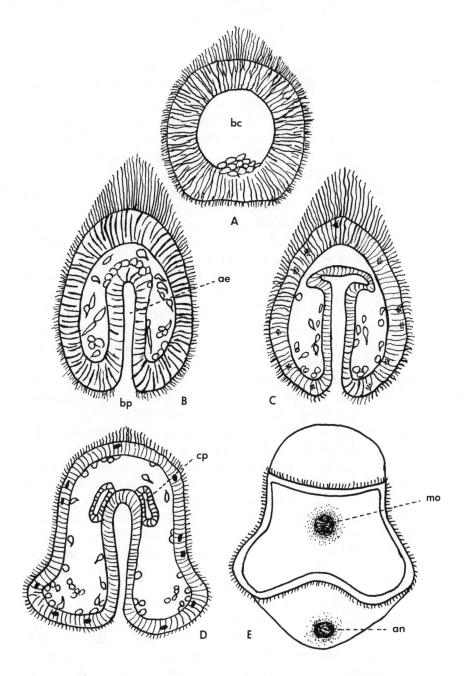

FIG. 5.2. Stages in the early development of an echinoderm. (a) Blastula; (b) gastrula; (c,d) formation of coelomic pouches from archenteron; (f) early bilaterally symmetrical larva. a–d in section; e in ventral view. ae, archenteron; an, anus; bc, blastocoel; bp, blastopore; cp, coelomic pouch; mo, mouth. (Adapted from Pruho, 1888.)

In the Asteromorpha the larva develops into a "bipinnaria" (Fig. 5.3a), so called because the cilia are in two separate bands. Some starfish metamorphose from the planktonic bipinnaria larva, but in many the bipinnaria develops into a "brachiolaria," the name referring to "short arms" with organs of attachment. This later larval form settles before metamorphosis is completed.

Most members of both the Ophiuromorpha and the Echinomorpha have a "pluteus" larva (Fig. 5.3b, c). Latin dictionaries usually translate "pluteus" as a mobile shelter, and if the orientation of the larva is inverted from that shown here it looks rather like a miniature wigwam. The Latin name, however, can also mean a desk, and Johannes Müller, who first named such larvae, probably had this definition in mind, for he compared their shape to that of an artist's easel. This type of larva has slender ciliated arms supported by an endoskeleton of calcareous rods, a unique combination of characters. In some species the arms are spread out, like the supports of a circular wigwam, but very often the whole body is dorsoventrally flattened. Pluteus larvae of the Ophiuromorpha are known as ophioplutei, and in them the outermost pair of arms tend to be the longest, which they rarely are in echinoplutei, the larvae of the Echinomorpha. The number of arms can vary between species in both groups, and it is not always easy to say whether a larva is an ophiopluteus or an echinopluteus, although the corresponding adults are so markedly different. Some ophiuromorphs and a smaller number of echinomorphs undergo direct development, with no free larval stage, but there are also a few ophiuromorphs with a planktonic larva very different from a pluteus, known as a "doliolaria." This larval form also occurs in the Holothuromorpha and Crinomorpha, and it is described below.

In those Holothuromorpha with a planktonic larva the gastrula develops into an "auricularia" (Fig. 5.3d), so called because the pattern of the convoluted band of cilia bears some resemblance to a human ear. The continuous band of cilia that propels this stage distinguishes the larva from a bipinnaria, with two bands, but as development proceeds the au-

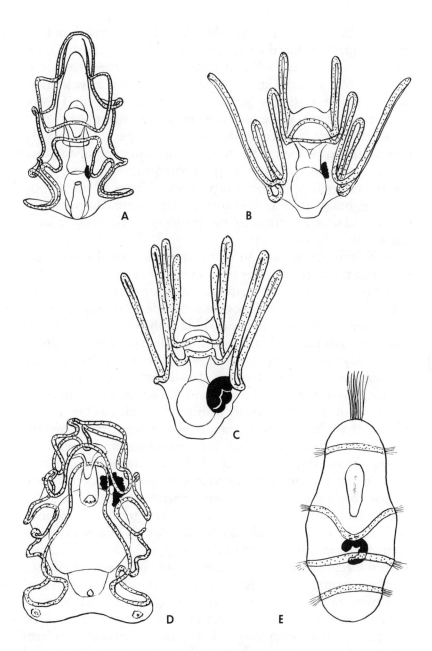

FIG. 5.3. Larvae of the Echinodermata. (a) Bipinnaria (Asteromorpha); (b) pluteus (Ophiuromorpha); (c) pluteus (Echinomorpha); (d) auricularia (Holothuromorpha); (e) doliolaria (Crinomorpha). A doliolaria also occurs, after the auricularia, in the Holothuromorpha, and as the only larva in some Ophiuromorpha.
Developing juveniles within the larvae are shown black. (a–c redrawn from Mortensen, 1931; d,e redrawn from Chadwick, 1914.)

ricularia becomes more cylindrical, and the single band of cilia becomes replaced by a number of distinct rings. There are usually five such ciliated rings, but in some cases only three or four, and the larva propelled by them is now known as a "doliolaria" (Fig. 5.3e), meaning barrel shaped. This is the form that gives rise to the juvenile sea cucumber.

Development is direct in the majority of species of the Crinimorpha, but some species have a planktonic larva, and this also is a doliolaria. Like the ophiuromorph doliolaria, it relies on internal yolk for nourishment. There is no mouth opening, but a well-marked surface depression, the vestibule or stomodaeum, surrounds the position of the mouth in a feeding doliolaria.

Doliolaria larvae are, in many respects, radially symmetrical, but the coelom arises from lateral sacs, as in other echinoderm larvae, and the vestibule, with or without a mouth, marks the ventral side. All other echinoderm larvae are quite clearly bilaterally symmetrical. This is the only phylum with bilaterally symmetrical larvae and radially symmetrical adults, and biologists have been asking why ever since the concept of evolution became accepted. Until now, the only explanation that has been put forward theorizes that the early echinoderms were bilaterally symmetrical both as larvae and adults, that the adults adopted radial symmetry as an adaptation to sessile life, but the free-swimming larvae remained bilaterally symmetrical. This explanation implies that the bilaterally symmetrical ancestors of the echinoderms were all soft bodied and left no fossil remains, for, although the fossil record of the echinoderms is very good and extends back to the Lower Cambrian, it includes no bilaterally symmetrical forms. Some asymmetrical fossils, whose affiliation with the echinoderms is debatable, have been described, but a clear distinction should be drawn between asymmetrical and bilaterally symmetrical, and the oldest known echinoderms were clearly radially symmetrical (Paul, 1979, and see Chapter 9). I consider the evidence for bilateral echinoderm ancestors unconvincing, but, if they ever existed, why did they become extinct while their radial relatives survived and spread? Bilateral symmetry is generally considered to have advantages over radial symmetry, particularly in motile animals, and although many of the earliest known echinoderms were sessile, others were not. Even if it is supposed that there was a period in the pre-Cambrian evolution of the echinoderms when all members of the phylum were sessile, it is not clear why this life-style should have led to the elimination of bilateral symmetry in this group by the early Cambrian when other sessile groups such as the Bryozoa and the Brachiopoda have retained clear bilateral symmetry to the present day.

Smiley (1986, 1988) suggested that the partial bilateral symmetry of adult holothurians has been retained from the postulated bilateral ancestors of the whole phylum, but practically all palaeontologists have regarded the bilateral symmetry of this group as a relatively recent development. Raff et al. (1988) used molecular analysis to study the relatedness of the different echinoderm classes, and although they do not entirely rule out the possibility that the holothurians separated from all other extant classes very early in the evolution of the phylum, their preferred phylogenetic tree shows the separation of the holothurian and echinomorph lines as the last major branch in the evolution of the echinoderms. Paul and Smith (1984) and Smith (1984, 1988) reached the same conclusion from palaeontological evidence. Raff et al. also interpret their results as indicating that the holothurians are evolving more quickly than other echinoderm lineages. I would suggest that the bilateral tendencies of modern adult holothurians are explicable as an example of neoteny or paedomorphosis, the retention of a larval character by the adult. It may be recalled that the late larva of the holothurians is a doliolaria, with a tendency toward radial symmetry. This form of symmetry may be regarded as an adult feature making its first appearance in the late larva. In relation to their ancestors, modern holothurians seem not only to have accelerated the ontogenetic appearance of a degree of radial symmetry, so that it now makes its first appearance in the late larva, but also to have retarded the complete disappearance of bilateral symmetry, so that it now persists throughout adult life. The group, therefore, provides examples of both positive and negative heterochrony in development, but these are likely to be comparatively recent features in this line, which, as we have just seen, appears to be evolving relatively rapidly.

The theory that radially symmetrical adult echinoderms evolved from bilateral ancestors seems to raise more difficulties than it solves, but I have found little serious criticism of it. H. Baraclough Fell, the eminent New Zealand authority on the phylum, has repeatedly asserted that echinoderm larvae give totally misleading clues on relationships within the phylum and on the affinities of echinoderms to other phyla, but even he seems to have accepted that the bilateral larvae of modern echinoderms must have come from ancestral echinoderms which were themselves bilateral both as larvae and adults. I agree that the larvae must have come from forms that were bilateral both as larvae and adults, but I suggest that these forms were not echinoderms and belonged to an entirely different evolutionary line. The dromioid crustaceans, considered in the previous chapter, provide an example of a superfamily that might be descended from a line which had lost its zoeal phase in development until one of its members acquired a replacement larva from another group of crustaceans. The echinoderms seem to provide a broadly comparable case at the phylum level, although here there is no reason to postulate the loss of an earlier larva. I suggest that modern echinoderms evolved from forms that were radially symmetrical throughout life and had no planktonic larvae. As we shall see in Chapter 7, at least one modern species exhibits this type of development, which I regard as ancestral. I further suggest that one of these early radial echinoderms with direct development then acquired a bilaterally symmetrical larval form from another phylum by horizontal genetic transfer. Whereas a minority of modern echinoderms with bilaterally symmetrical larvae are directly descended from this ancestor, the majority acquired their larvae by horizontal genetic transfer within the phylum. This would explain why there are no bilaterally symmetrical fossil echinoderms, and it would remove the necessity of explaining how and why a group of bilaterally symmetrical animals gave rise to a successful line of radially symmetrical descendants before themselves becoming extinct.

It is, of course, relevant to ask how a larval form might be transferred from one animal to an unrelated one, and cross-fertilization is considered as a possible mechanism in a later chapter. Before that, however, I shall point out many more paradoxes that can be solved by assuming that, from time to time, such larval transfers have taken place, and as the weight of evidence accumulates I hope the theory of horizontal transfer of larval form will become more generally acceptable. Next, let us consider further the suggestion that, at some time in their history, the echinoderms acquired their bilateral larvae from another phylum.

6
Echinoderms: Relationships with Other Phyla

Main features of phylogenetic tree of animal kingdom generally agreed by early 20th century but now disputed—Echinoderms and hemichordates usually regarded as outgrowths of same major branch of tree, but resemblances restricted to larvae—Similar larvae of these phyla point to common origin but only of larvae—Suggested that original echinoderms had no larvae—Larval form transferred from ancestor of modern hemichordates after existing classes of echinoderms had evolved; subsequent transfers between representatives of different classes.

Throughout the first half of the nineteenth century the publications of Jean Baptiste Lamarck and Léopold Chrétien Fréderic Dagobert Cuvier (known as Georges) were generally accepted as standard works on systematic zoology, and both these illustrious French biologists had grouped the echinoderms with medusae and corals in the Radiata. The members of the Radiata had virtually nothing in common except radial symmetry, and, in the middle of the century, Frey and Leuckart (1847) and Leuckart (1854) pointed this out, noting that the echinoderms, with three distinct body layers and a well-developed body cavity, were at a "higher" grade of organization than medusae and other coelenterates, with only two layers and no coelom. Embryological studies on a number of groups enabled Huxley (1875) to associate the Echinodermata with the Chaetognatha and Enteropneusta as enterocoelous deuterostomes, and this link was reinforced when Metschnikoff (1881) drew attention to the resemblances between the larvae of echinoderms and enteropneusts. The Enteropneusta were grouped with the Pterobranchia in the phylum Hemichordata by Harmer (1887), and the Chordata were added to the enterocoelous deuterostomes by Götte (1902) to give the association of phyla widely recognised today either as a superphylum or as outgrowths of the same major branch of the phylogenetic tree (Fig. 3.1).

This phylogenetic tree has been a familiar part of the zoological landscape throughout my lifetime, and old trees, whether phylogenetic or botanical, undeniably have their attractions. They should, however, be inspected periodically to see if they are still healthy, and in some cases the only responsible course of action is to cut them down. Perhaps the reader should be warned that, if my views are accepted, little but the stump of this tree will remain intact. At this stage it is necessary to question the grouping of the echinoderms with the hemichordates that the tree depicts.

The phylum Hemichordata was so named because its members were considered to be half way to becoming chordates. It is also sometimes known as the Stomochordata and is made up of two main classes, the Enteropneusta and the Pterobranchia. Passing mention may also be made of a third class, which is sometimes recognised for the genus *Planctosphaera;* its members are probably giant larvae, but the adult form is unknown. The name Enteropneusta may be translated as "gut lungs," and refers to the fact that the animals of this group have a pharynx provided with gill slits. This feature links the group with the chordates. Enteropneusts are often called "acorn worms" because of the acorn-shaped proboscis that makes up the anterior part of the body and the worm-like trunk that makes up the posterior section and can account for 90% of the total length. Between these anterior and posterior sections is a relatively short region of the body, the collar. The body is, therefore, trimerous, consisting of three distinct regions (Fig. 6.1a). Most of the known species live in tubes in marine sediments, but in recent years some have been found draped over rocks in the proximity of hydrothermal vents, local areas of volcanic activity on the ocean floor. *Balanoglossus* is the genus mentioned in most textbooks of invertebrate zoology, and many of the generic names contain either the suffix "-glossus" or the prefix

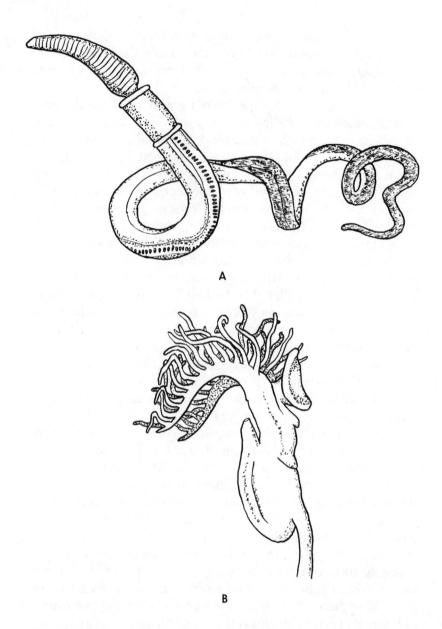

A

B

FIG. 6.1. Adult and larval hemichordates. (a) Adult *Dolichoglossus* (Enteropneusta); (b) adult *Rhabdopleura* (Pterobranchia); (c–f) metamorphosis of an enteropneust; (c) tornaria larva; (d,e) metamorphosing larvae; (f) settled juvenile. (a,b redrawn from Borradaile et al., 1935; c–f redrawn from Hyman, 1959.)

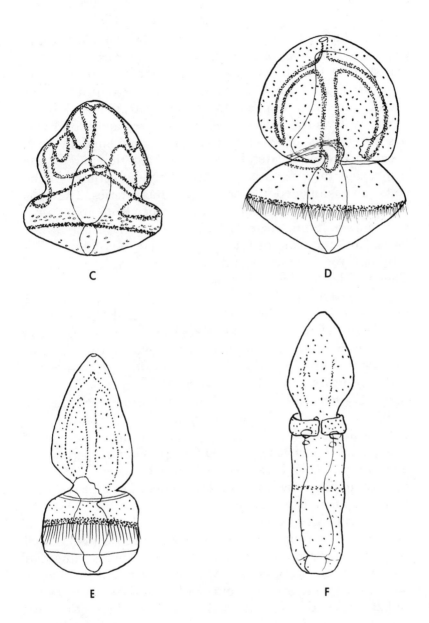

C

D

E

F

"glosso-," literally meaning a "tongue," but referring to the proboscis. The body is approximately cylindrical, but bilateral symmetry is shown in the arrangement of most of the principal organs. Thus the mouth is on the ventral side of the anterior collar, opposite the dorsal heart; the trunk bears dorsal and ventral ridges containing nerves and blood vessels; the gill slits, of which there can be several hundred pairs, occur in two dorsolateral rows in the anterior part of the trunk, and the coelom of the trunk and sometimes of the collar also is divided by dorsoventral mesenteries. The anus is at the posterior extremity. Lengths range from about 2 cm to over 2 m. Some members of the group have planktonic larvae, and these are of very great interest in relation to the affinities of echinoderms and the wider theme of this book. They will be considered below.

The body of the Pterobranchia (Fig. 6.1b) also consists of three regions, as in the Enteropneusta, but the regions are less clearly delimited. The class takes its name, meaning "wing gills," from a series of paired arms that extend from the dorsal side of the collar: there can be from one to nine pairs, and each bears a double row of ciliated tentacles. The proboscis is often shield shaped, and the intestine curves forward, inside the sac-like trunk, to open at a dorsal anal papilla near the collar. There is never more than one pair of pharyngeal slits. The body is stalked, providing a permanent attachment in colonial forms, temporary in others. The length of each animal ranges from a few millimetres to a few centimetres. There is no larval stage.

The trimerous, bilaterally symmetrical body of adult hemichordates is in such marked contrast to the pentaradial body of adult echinoderms that the only obvious thing that animals of the two groups have in common is that they are all coelomate metazoans. Some early echinoderms were triradial, rather than pentaradial, and a few biologists have suggested that this is a link with the trimerous hemichordates, but the triradial echinoderms were no more like hemichordates than the pentaradial forms (see Chapter 8). Others have suggested that the tentacles on the arms of pterobranchs are homologous to the tube-feet of echinoderms, but the tentaculated arms of pterobranchs form an organ very similar to the lophophores of brachiopods, phoronids, and bryozoans and suggest that the true affinities of the hemichordates are with these other lophophorate groups. If the larvae of the enteropneusts were unknown it would never have occurred to anyone to suggest that the hemichordates and the echinoderms had any common ancestry later than the original stock of the metazoa.

Some enteropneusts have no free-living larva, but many pass through a planktonic "tornaria" stage (Fig. 6.1c), so called because it spins as it swims. Tornarias bear such resemblance to echinoderm larvae that the authors of all the early descriptions ascribed them to that phylum, usually to the Asteromorpha. There is a ring of strong cilia round the posterior part of the body and a convoluted ciliated band winds its way over much of the remainder of the body surface. The general resemblance to a bipinnaria or an auricularia is further strengthened by the method of formation of the coelom. In some species an anterior vesicle with a posterior projection on either side becomes divided off from the archenteron, then each of the posterior projections separates from the anterior vesicle and divides into two, thus producing a single anterior protocoel sac and paired mesocoel and metacoel sacs. In other species the paired mesocoel and metacoel sacs are never connected to the protocoel but bud off independently from the archenteron. The result, however, is the same in both cases, and the larva is clearly enterocoelous. It also resembles an echinoderm larva in being deuterostomatous, for the anterior end of the archenteron bends ventrally to connect with a surface depression and thus form a new mouth.

As far as the pattern of ciliated bands is concerned, a tornaria bears much greater resemblance to an auricularia or a bipinnaria than to a doliolaria or a pluteus. All these larval forms can, however, be derived from an ancestral type resembling a tornaria. A doliolaria does indeed develop from an auricularia during the ontogeny of a holothurian, and a pluteus may be regarded as having evolved from a form similar to a bipinnaria with the addition of calcareous rods to support the ciliated projections.

The present widely accepted theory is that echinoderms and hemichordates must have had a common ancestor that gave each group its basic larval form. I subscribe to the minority view that the adult morphology of the two phyla is quite incompatable with recent common ancestry, a view that H.B. Fell put forward in 1948 and has repeated several times since. Although agreeing with Professor Fell's interpretation of the evidence from the adults, I do not agree with his dismissal of the evidence from the larvae. His conclusion was that ontogeny of echinoderms (the successive stages through which they develop) gives no clues to their phylogeny (their ancestry and relationships). The opposite view, expressed by Darwin and many others, is that, for animals in general, the ontogeny of a group holds irrefutable evidence of its phylogeny. That is not to say, as Haeckel did well over a century ago, that embryos and larvae are miniatures of ancestral adults: a theory encapsulated in the aphorism "ontogeny recapitulates phylogeny." It *is* to say that larvae and patterns of development have themselves evolved, and that phylogeny can be deduced from them just as surely as from adult morphology and sometimes more surely.

The theory that I am now advancing seeks to reconcile the general view that there is a connection between ontogeny and phylogeny with Fell's view that, in the case of echinoderms, there is no connection between the larval development and the evolution of the group. The morphological characters of embryos and larvae undoubtedly reflect their own ancestry, but if these characters, or the genes that determine them, can sometimes be transferred from one group to another then the ancestry of embryos and larvae need not always be the same as that of the adults with which they are now associated. Echinoderms and hemichordates, I suggest, evolved quite independently from the original stem of the Metazoa. The life history of the earliest hemichordates may well have included a ciliated planktonic larva, and from this the tornaria evolved. I postulate, however, that, unlike the

early hemichordates, the life history of the earliest echinoderms probably included no planktonic phase. It was not until comparatively late in the history of this phylum that one
of its members acquired a tornaria-like larva from an ancestor of the modern enterpneusts. Over many millions of years,
the genetic prescription for this larval form, or for other larvae evolved from it, then spread to other echinoderms, so
that the ancestry of all modern echinoderm larvae can be
traced back to the original horizontal transfer from a hemichordate, a very distantly related animal.

So far, I have drawn my evidence on the relationships of
echinoderms and enteropneusts from the morphology of the
adults and larvae of the two groups, and I claim that this
evidence is fully consistent with the suggestion that a larval
form that originated in the latter group was later transferred
to the former. More evidence in support of this view comes
from a consideration of the way in which the larva develops
into the juvenile and hence the adult in each of these two
groups. This process of metamorphosis will be considered
next.

7
Echinoderms: Metamorphosis

Hemichordate larva metamorphoses by differential growth; most larval organs and tissues contribute to body of juvenile—Juvenile echinoderm develops from cells surrounding one or more coelomic sacs of larva; much of larva discarded and any larval tissues retained probably undergo histolysis—Misleading to regard adult echinoderms as deuterostomes—Starfish *Luidia* shows remarkable independence between juvenile and larva—'Kirk's ophiuromorph' develops directly as a schizocoelous protostome: suggested original method of development of echinoderms—Differences in methods of metamorphosis in different classes of echinoderms and within Ophiuromorpha

An animal species, as Darwin reminded his readers, comprises all phases in development from egg to adult, and eggs, embryos, larvae, juveniles, and adults have all evolved. During evolution, each of these stages in development must have remained fully functional, for survival clearly implies the survival of them all, and as the various phases in development have changed, the method of changing from one to another must also have been modified. If all modern species are solely the result of descent with modification from species to species, then the method of metamorphosis from one phase to another must have evolved with the phases concerned. As we shall see, metamorphosis in echinoderms is a totally different process from that in enteropneusts, even though it starts from similar larvae, and it must be questioned whether the two processes could have evolved from a common ancestral process.

An adult *Balanoglossus* is very different in form and way of life from its tornaria larva, but the transformation is practically all accomplished by cell division and differentiation. The larva develops a waist, and the portion in front of this constriction becomes the acorn-shaped proboscis. The body behind the restriction elongates considerably and differentiates into the collar and trunk as the bands of cilia become replaced by more uniform ciliation and the young animal settles (Fig. 6.1c–f). The single protocoel sac provides the undivided proboscis coelom, and the paired mesocoel and metacoel sacs expand to form the body cavities of the collar and trunk, respectively, the left and right components of the collar fusing in some species but those of the trunk always remaining separated by mesenteries. The orientation of the juvenile, with respect to anterior and posterior, dorsal and ventral, left and right, is the same as that of the larva, and the mouth, alimentary canal, and anus of the tornaria develop into the same organs of the settled juvenile. Virtually the only parts of the larva to be discarded during metamorphosis are its natatory cilia.

A tornaria larva can certainly be said to "develop into" a juvenile enteropneust, but no echinoderm larva can be said to "develop into" a juvenile echinoderm. It is rather that the juvenile echinoderm develops within the larva, almost as a separate animal.

Some features of the metamorphosis of echinoderms are much the same whether the larva is a bipinnaria, brachiolaria, pluteus, auricularia, or doliolaria. The larva will usually go through a stage with three pairs of coelomic sacs, as in the pluteus of *Ophiocomina* shown in Figure 7.1, but one or more of the sacs may fail to appear or have only a very brief existence. In the literature on echinoderm development they are often called axocoel, hydrocoel, and somatocoel sacs, but the terms protocoel, mesocoel, and metacoel used in the development of the Enteropneusta are equally applicable. The left mesocoel sac is usually bigger than its right counterpart, and as it grows still larger it develops five lobes, which soon become arranged radially around the sac. This is the first indication of the pentaradial symmetry of the adult. It should be noted that the arms of adult starfish or brittle-stars develop quite independently of the lobes or arms of their respective larvae, and other features that emphasize the independence of the larvae and juveniles will be mentioned below.

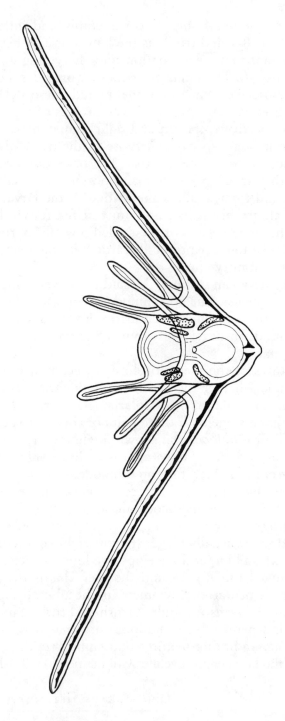

FIG. 7.1. A fully developed pluteus larva of *Ophiocomina nigra* (Ophiuromorpha), showing the coelomic sacs (stippled). The left mesocoel sac has started to develop the five primary podia of the juvenile. (Redrawn from MacBride, 1914.)

The five-lobed rudiment around the left mesocoel sac will grow to form most of the radial structures of the developing juvenile, but it will receive some juvenile radial components from one or occasionally both metacoel sacs. In the case of a developing holothuromorph the juvenile grows around the alimentary canal of the doliolaria larva and incorporates it, without change of function, into the juvenile body. In this group the orientation of the long axis of the advanced juvenile comes to coincide with that of the doliolaria, with the juvenile mouth at the anterior end of the larva. In those Crinomorpha that pass through a doliolaria stage the eventual orientation of the juvenile is exactly the opposite of that in the Holothuromorpha. The base of the stalk of the developing juvenile lies at the anterior end of the larva and the main juvenile body at the posterior end, with the mouth facing posteriorly. In the Ophiuromorpha the juvenile migrates to surround the esophagus of the pluteus, so that the eventual oral orientation of the juvenile and of the larva is the same, anteroventral with respect to the larval body. In both the Asteromorpha and the Echinomorpha the juvenile mouth normally faces to the left side of the larva until the time of settlement. The amount of larval alimentary canal incorporated into the juvenile varies considerably from class to class, being greatest in the Holothuromorpha and least in the Echinomorpha. In the Holothuromorpha the juvenile gut may develop directly from the larval gut by differential growth of the cells, but this requires verification. In all other classes with larvae, any parts of the larval gut that are incorporated into the juvenile are first broken down, and the cells lose their orientations and special functions before being redifferentiated to the juvenile condition. In all cases the preoral lobe of the larva and its coelom, virtually all the epidermis and any skeletal rods degenerate and are used as food by the juvenile or are discarded by it (Chia and Burke, 1978; Strathmann, 1978).

The ectoderm, endoderm, mesoderm, and coelom of an adult enteropneust are the direct descendants of the corresponding parts of the larva, and the larval mouth becomes the adult mouth, so that the adult, like the larva, is an enterocoelous deuterostome. Most adult echinoderms may also be said to be enterocoelous because the various parts of the adult coelom are developed from some of the larval enterocoelic pouches, although the trimerous larval condition is not retained. Enterocoelous development in this phylum, however, is not directly comparable with that in the Enteropneusta. In the Echinodermata growth of the mesenchyme that surrounds the coelomic pouches determines the shape and symmetry of the juvenile and hence the adult, thus giving this tissue a role it does not have in the Enteropneusta or, indeed, in any other group of animals. The larval cells of this mesenchyme are also probably the only ones to produce adult cells by direct descent. It has been established in an asteromorph and in an echinomorph that no larval ectodermal or endodermal cells are incorporated into the juvenile until they have undergone histolysis (and in the case of ectoderm cells cytolysis also) and redifferentiation, and this may well be the case throughout the phylum (Chia and Burke, 1978). The direct development of a brittle-star, whose coelom is a typical schizocoel, will be described below, but although the coelom of the majority of known echinoderms may be traced back to a larval enterocoel, it should be noted that the adult coelom develops very differently from that in other enterocoelous phyla.

A reassessment should also be made of the oft-repeated statement that the echinoderms are deuterostomes, that is, the mouth is an aperture quite independent of the blastopore of the gastrula. That all echinoderm *larvae* are deuterostomes is undisputed. In those holothuromorphs and ophiuromorphs whose life history includes a larval phase, the juvenile first forms with an orientation quite different from that of the larva, but it migrates to take over the larval mouth, so most juveniles and adults of these classes may perhaps also be referred to as deuterostomes. On the other hand, the schizocoelous ophiuromorph mentioned in the previous paragraph is also a protostome. But returning to cases of development involving a larva, the larval mouth does not become the adult mouth in other classes. In the development of the crinomorph doliolaria, the blastopore closes early, and the larva has no functional mouth or anus. The developing juvenile, within the doliolaria, takes up a position with its mouth facing the original position of the blastopore, at the posterior end of the larval body. Does this make the crinomorphs protostomes? In asteromorphs and echinomorphs the juvenile mouth is a new structure, quite independent of the blastopore or the larval mouth, so that, although the larva is a deuterostome, this term has no meaning in relation to the juvenile and adult. Larval enteropneusts and echinoderms can certainly be classed together as enterocoelous deuterostomes, but although these terms are also applicable to adult enteropneusts it is quite misleading to apply them to adult echinoderms. If we restrict ourselves to conventional evolutionary theory this seems to be an almost inexplicable anomaly, but not if we depart from conventional theory.

If all stages in the development of the Enteropneusta and the Echinodermata have evolved by descent with modification from a common ancestor, then the two utterly different processes of metamorphosis that occur in the two groups must have evolved from the process practiced by that ancestor. Let us assume, for the moment, that the metamorphosis of this ancestor was similar to that of modern enteropneusts, and most of the larval tissues and organs were utilized by the developing juvenile. Conventional evolutionary theory would presumably postulate that, to get to the echinoderm condition, there must have been methods of metamorphosis utilizing progressively less of the larva until the cells surrounding a single coelomic sac determined the symmetry and orientation of the juvenile. The difficulties appear to be at least as great if we assume that the common ancestor metamorphosed by the echinoderm method. Not only are there no known intermediate types of metamorphosis, but I cannot envisage any. My proffered explanation of the bizarre method of metamorphosis practiced by echinoderms (and of all the anomalous differences between the larvae and adults of that phylum) depends on the unorthodox assumption that the original echinoderm larva was a late insertion into a preexisting life history. To change from the larva to the juvenile, the developing animal would have had to change from one phylum to another, and this drastic change required drastic methods. The configurations of cells surrounding the coelomic sacs were probably the only ones in the newly acquired larva that bore any resemblance to groupings of cells that occurred during the development of other contemporary echinoderms, with no larvae. It was, I suggest, differential growth of these larval mesenchyme cells that produced the first juvenile echinoderm from a tornaria-like larva, and the rest of the larva was useful chiefly as a source of food and transport.

Those who are familiar with the life histories of insects will, no doubt, point out that many members of that group, too, undergo a quite drastic form of metamorphosis in the pupal stage, involving the histolysis of certain tissues, particularly the digestive system and muscles, and their regeneration from localised groups of cells known as imaginal discs. Other insects, however, develop very gradually from the larva to the adult, without a pupa or other inactive period, and, in those insects with pupae, there is very great variation in the length of the pupal phase and the changes that take place during it. The insects, therefore, differ from the echinoderms in that a whole series of intermediate conditions is known, linking gradual development with drastic metamorphosis. They also differ in retaining the same orientation of the body throughout development, and, although certain organs undergo histolysis and histogenesis, the dorsal blood vessel and the central nervous system are little affected and pursue an uninterrupted course of differentiation (Imms, 1946). Even the most extreme cases of metamorphosis in insects may be examples of abbreviation of development, in which what were originally a number of larval instars have been telescoped into one pupal instar. Perhaps, however, the caterpillar may be regarded as a phase in development transferred from an onychophoran or a myriopod. It could help to explain the occurrence of this type of larva not only in all lepidopterans but also in some hymenopterans. With the questionable exception of metamorphosis from caterpillars, it is not difficult to envisage that the various forms of insect metamorphosis might have evolved by the accumulation of a series of small changes. I am, however, quite unable to conceive how the metamorphosis of echinoderms could have evolved in a corresponding way.

There is always a marked degree of independence between a juvenile and a larval echinoderm, and a striking example of this occurs in the development of the starfish *Luidia sarsi*. In many of the Asteromorpha the planktonic bipinnaria larva is succeeded by a brachiolaria. This larva has adhesive cells and a sucker, and it anchors itself to the bottom before degenerating and freeing the benthic juvenile. In *Luidia* and several other genera there is no brachiolaria stage, and the bipinnaria goes on swimming until the juvenile is ready for settlement. In most cases the larval body is then resorbed and acts as nourishment for the juvenile; in *L. sarsi*, however, the large bipinnaria shows no sign of degeneration as the juvenile separates, and one case is recorded of the larva continuing to swim actively for a further 3 months after the juvenile had separated and crawled away (Tattersall and Sheppard, 1934) (Fig. 7.2). Certainly this larva does not "develop into" the juvenile.

In larvae of all classes of echinoderms the left mesocoel sac is usually larger, and indeed in crinomorphs and holothuromorphs the right mesocoel sac never appears. Normally the cells surrounding the left sac produce the five primary podia of the juvenile rudiment, and the right sac, if present, disappears. Occasionally, however, the right mesocoel sac of asteromorphs, echinomorphs, and ophiuromorphs is the larger, and this then develops the five primary podia and the left sac disappears. Much more rarely the left and right mesocoel sacs are of similar size, and then twin juveniles may develop. Such a case in the development of the common British sea-urchin, *Echinus esculentus*, was described by MacBride (1911) (Fig. 7.3). It will be seen from the drawing of the earlier stage that although the juveniles

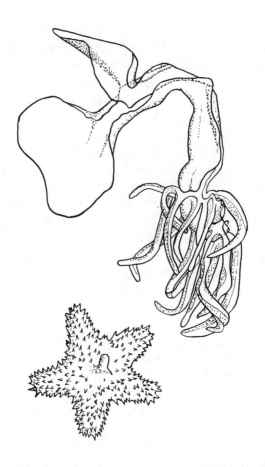

FIG. 7.2. Swimming bipinnaria larva and settled juvenile of the starfish *Luidia sarsi*, just after their separation. (Redrawn from Tattersall and Sheppard, 1934.)

had then reached a fairly advanced stage of development, they apparently had not incorporated any parts of the larval epidermis or stomach. The later stage shows the twin juveniles almost ready to go their separate ways as benthic urchins. This case again emphasises the independence of the larva and the juvenile (or juveniles), and it is fully consistent with the suggestion that the only larval feature necessary for the development of a juvenile is one or more coelomic sacs of suitable size and shape.

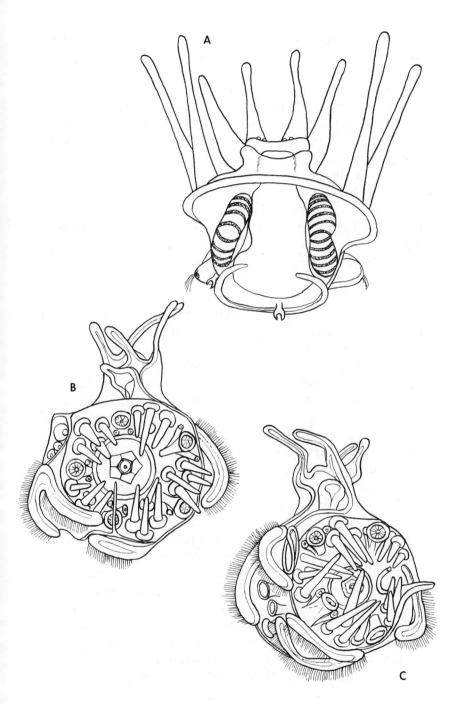

FIG. 7.3. Twin juveniles of the sea-urchin *Echinus esculentus* in the same pluteus larva. The upper drawing (in ventral view) shows an earlier stage of development than the two lower drawings (from left and right). (Redrawn from MacBride, 1911.)

The thesis was put forward earlier (at the end of Chapter 5) that modern radially symmetrical echinoderms with bilateral larvae are descended from ancestors that had no larvae and were radially symmetrical throughout life. We know that the early life history of all existing triploblastic animals includes a gastrula stage (described in Chapter 2), usually the last radially symmetrical stage in their development. If, however, the animal is to become pentaradial rather than bilateral we might expect to see that form of symmetry developing in the late gastrula. I suggest that the appearance of the five primary podia as outgrowths of the left mesocoel of modern echinoderm larvae mirrors similar development from a gastrula in ancestral species.

When the life history of an animal includes a larval phase as well as juvenile and adult phases there must be a genetic mechanism to keep these phases in their right order: larva, juvenile, adult. If, as here suggested, an animal with no larval phase then acquires one from another species, this can be envisaged as resulting from the transfer of genetic material that carries not only the prescription for a larval form but also the genetic instruction to give the larva priority over the juvenile and adult. Applying this to the postulated echinoderm that acquired a bilaterally symmetrical larva from another group, when it reached the gastrula stage it would have had the genetic prescriptions to make both a bilateral larva and a radial juvenile, but the larva would have taken priority. When the larva was well developed, the genetic instructions for the development of the juvenile would then have come into force. In all earlier generations of echinoderms these instructions would have led to the gastrula developing pentaradial structures, but in the late larva they would have meaning only for groups of cells whose size and shape were comparable to those of a late gastrula. The cells surrounding one or more of the larger coelomic sacs provided these conditions, as they do today in modern echinoderm larvae.

An animal may invest its reproductive assets in many offspring that disperse widely as planktonic larvae, in few offspring with food and shelter provided by the parent, or in a compromise between these two extremes. Wide dispersal might enable an animal to colonise new habitats or new areas, and it can be of particular advantage in times of climatic change as it increases the chance for some of the progeny to reach suitable conditions for survival. When conditions are predictable, however, it is less wasteful to make better provision for relatively few, well-adapted offspring. This may be achieved by shortening larval life (abbreviated development) or by eliminating it altogether, so that the embryo develops directly to a juvenile, with no intervening larval phase (direct development). With fewer offspring the female can provide food, in the form of yolk, and perhaps physical protection also. This has particular advantages when conditions are harsh, as in polar seas, and examples of abbreviated and direct development are more common in such conditions, but by no means restricted to them. All the surviving classes of echinoderms contain examples of direct development, but in nearly all these the juvenile develops in much the same way as it would from a planktonic larva: the embryo produces enterocoelic sacs, similar to those of a free larva, and the juvenile grows from the cells surrounding these sacs. This suggests that such species have evolved from others that had larvae, and that their direct development is therefore secondary. We know that the history of the echinoderms has included periods of mass extinction, and it may be supposed that the bulk of the survivors were those with planktonic larvae. At one time I assumed that all modern echinoderms had descended from others with larvae, and that all those postulated forms that remained completely radial throughout life were now extinct. I was, however, delighted to find that my assumption was wrong: there are extant species that develop without any trace of a bilateral larva, a bilateral coelom, or any other bilateral feature.

The development of a New Zealand brittle-star in this category was described in detail by H.B. Fell in 1941, then a research fellow at Wellington. A preliminary account of the development of the same species had been published in 1916 by Professor H.B. Kirk, but both Kirk and Fell collected fertilised eggs attached to stones at extreme low water spring tides, without being able to find out which female had laid them, and the identity of the species remains uncertain to this day. Fell referred to it as "Kirk's ophiuroid," which in the nomenclature I have adopted becomes "Kirk's ophiuromorph." He reared specimens at 13.5–15°C in small dishes through which seawater was slowly circulated. The egg, 0.5 mm in diameter, after repeated divisions produces a blastula with a rather small blastocoel (Fig. 7.4a). Invagination and overgrowth of cells lead to the formation of a gastrula with a very small blastocoel (Fig. 7.4b), and further invagination and cell growth continue until both the blastocoel and the archenteron are filled with cells, although the blastopore remains as a depression in the surface. After 6 or 7 days, five grooves radiate out from the blastopore and the embryo assumes a pentagonal shape, with the grooves running toward the midpoints of the sides. By about the tenth day the blastopore depression is deeper, and a ring of 10 equally spaced, rounded projections has developed just beyond the tips of the grooves, on either side of each major radius, and the "angles" of the pentagon are now rounded lobes (Fig. 7.4c, d). The 10 projections are developing podia, and during the next 3 days a split appears between the cells in the middle of each. Next a circular split develops in the mesenchyme about halfway between the podia and the centre, and the podial cavities extend inward to join with the circular cavity, thus providing the water vascular system of the developing animal. A further cavity grows inward from the blastopore to form the stomach, so that the blastopore becomes the mouth. Hatching takes place about 15 days from first cell division, then the main coelom arises from further splits in the mesenchyme, the podia elongate, and spines appear (Fig. 7.4e,f). Spiny arms develop, and 15 days from hatching these show signs of segmentation (Fig. 7.4g). A further pair of podia develops on the oral side of each arm segment, and by the time five or six segments have developed the young brittle-star has used up all its yolk and begins to feed on diatoms and other small algae. As in all Ophiuromorpha, there is no anus.

Kirk's ophiuromorph certainly proves that a pentaradial echinoderm can develop from a radially symmetrical gastrula, just as I postulated that the ancestors of all modern echinoderms did. Its importance, however, goes far beyond this. Its coelom develops from splits in the mesenchyme, and its mouth develops from the blastopore: it is a schizocoelous protostome and therefore in the sharpest of contrast to echinoderm larvae, which are all enterocoelous deuterostomes. If, as Hyman (1955) suggested, the development of Kirk's species represents an extreme case of suppression of the larval phase, how could it have evolved a new method of coelom formation and a new relationship between the blastopore and the mouth, and why should it? I regard it as much more probable that this species has retained the ancestral form of echinoderm development, which included not only the derivation of a radial adult from a radial gastrula but also schizocoely and protostomy. Fell (1968) suggested that brittlestars of the families Ophiomyxidae and Gorgonocephalidae probably develop in a similar manner, but these have not been studied in detail. I have already pointed out that there are difficulties in regarding adult echinoderms as enterocoelous deuterostomes, even when they develop from larvae to which these epithets undoubtedly apply. I hold that all the evidence is consistent with regarding the phylum as having evolved from radially symmetrical, schizocoelous protostomes with, in most cases, the later insertion of a bilaterally symmetrical, enterocoelous, deuterostomatous, larval phase. It will be most interesting to determine whether *Xyloplax,* the only known member of the Concentricyclomorpha, develops in what I regard as the ancestral method, as a schizocoelous protostome. It is known to develop directly, without a larval phase. The fact that it has no mouth or anus does not mean that it cannot be assessed as a protostome or a deuterostome, for the orientation of the body is clearly the same as in the Asteromorpha and Ophiuromorpha. If the

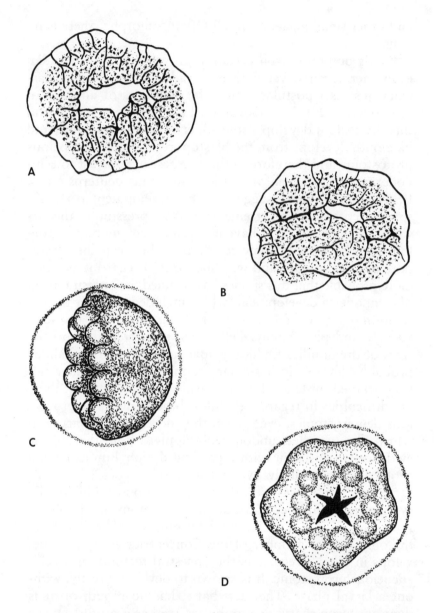

FIG. 7.4. Stages in the development of Kirk's ophiuromorph. (a) Section of blastula; (b) section of early gastrula (egg membrane omitted in a and b); (c,d) side and oral views of embryos with rudimentary podia; (e) newly emerged juvenile in oral view; (f) "asterina" stage in aboral view; (g) later juvenile with developing arms, aboral view. (Redrawn from Fell, 1941.)

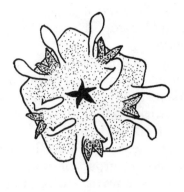

E

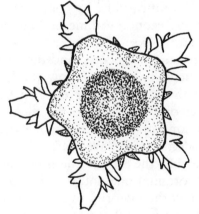

F

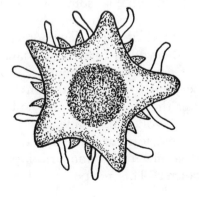

G

ventral end of the axis of the radially symmetrical body passes through the site of the blastopore, then the animal may be regarded as a protostome.

Even if it is accepted that genes governing early development can occasionally be transferred from one evolutionary line to another, it could still be argued that the form of development shown by Kirk's ophiuromorph is not necessarily the ancestral form for echinoderms. The genetic recipe for protostomy and schizocoely could have been transferred to Kirk's species from another group of animals. The form of development of Kirk's species is, however, so straightforward and economical that it is not difficult to envisage it as a product of Darwinian evolution, by the accumulation of relatively minor modifications. By contrast, we know of no straightforward and economical method of development, actual or hypothetical, for deriving a juvenile echinoderm through a process of enterocoely and deuterostomy, and it is extremely difficult to envisage how the known tortuous and wasteful methods could have evolved by conventional means. Whatever one's attitude to the theory of horizontal genetic transfer, one of the methods of early development known in echinoderms must have preceded the other. It seems highly probable that the simple method (shown by Kirk's species) was the original one and that the much more complex one (shown by the majority of echinoderms) was a later innovation, irrespective of its method of introduction. Also, one method of early development must have preceded the other in the Ophiuromorpha, so some ophiuromorphs (either the majority or the minority) must have undergone a radical change in their method of early development without any marked change in their adult form. The change in the form of development of echinoderms, therefore, could not have occurred in the very early history of the phylum, but it must have occurred after the Ophiuromorpha had become established as a separate evolutionary lineage. I suggest that the genetic recipe for a new form of embryonic development came with the recipe for a bilaterally symmetrical, planktonic larva, by horizontal transfer. This recipe was, I postulate, transferred between representatives of the five extant echinoderm classes with larvae, but the original transfer was from another phylum, the Hemichordata.

Another aspect of the metamorphosis of echinoderms is relevant to the debate on whether the life histories of the living members of the group are solely the result of descent with modification or whether the embryos and larvae of most species were later insertions. I refer to the differences between classes in the way that the juvenile develops. Mention has already been made of differences in the orientation of the juvenile within the larva, so that the juvenile mouth comes to face the anterior end of the larva in the holothuromorphs, the posterior end in the crinomorphs, anteroventrally in the ophiuromorphs, and laterally in the asteromorphs and echinomorphs. Under the conventional view there must have been an original orientation, and it is difficult to imagine why this should have been changed so much in the different groups, even if such changes are relatively easy. Although possible evolutionary changes in the orientation of the developing juvenile might present no insuperable problems, the recorded differences in the contribution made by the metacoel sacs to the developing juvenile, and hence to the adult, seem much more difficult to explain in terms of conventional theory. It has already been noted that one of the mesocoel sacs (usually the left) develops five lobes, which become the five primary podia, and in some cases one or other of the metacoel sacs can also become five lobed. Different parts of the metacoel may give rise to different adult structures, but the fate of the lobes is adequate to illustrate the differences from class to class (Hyman, 1955; Chia and Burke, 1978). The left metacoel forms five lobes in the Echinomorpha, Asteromorpha, and Ophiuromorpha. In the echinomorphs these lobes become the dental sacs from which the feeding apparatus known as Aristotle's lantern develops. In the asteromorphs the lobes bifurcate so that each projects into a pair of adjacent arms as parts of the radial hyponeural sinus system. In the ophiuromorphs the metacoel lobes also become parts of the system of hyponeural sinuses, but each lobe contributes to one arm only. In the Crinomorpha and Holothuromorpha the left metacoel sac forms no five-lobed structures, but the corresponding sac on the right side of crinomorphs develops five extensions, which become the chambered organs of the stalk. These differences between classes seem quite consistent with the suggestion that the larval phase was a relatively late addition to echinoderm life histories, arriving after the adult characters of the different classes were established. It is envisaged that only one or a very few individuals of each class successfully acquired a larva and that each of these devised its own method of metamorphosis. For each success there would almost certainly have been many failures, and probably many of these developed as far as the larva but failed to metamorphose to the juvenile. I return to the theme of the number of successes necessary for the hypothesis in Chapter 12, where I discuss the probable method of horizontal genetic transfer.

Differences in the method of metamorphosis are also to be found within the class Ophiuromorpha (Mortensen, 1931). Some of these differences apply to the larval arms, which may either be thrown off or resorbed at metamorphosis and which may change size, shape, and ciliation at this time; but I also wish to draw attention to differences in the growth of the mesocoel. In all families of the class that have a pluteus larva, the left mesocoel sac normally develops the five primary lobes and then migrates and elongates to encircle the larval esophagus. In some families it grows upward, then over the esophagus to encircle it counterclockwise (viewed from the mouth), whereas in others it grows under the esophagus and encircles it clockwise. Now under the conventional view changes in the method of metamorphosis may well have evolved in response to changes in larval form, adult form, or both, but if this were the explanation we might expect to see comparable variations in the method of metamorphosis among the Echinomorpha, which also have a pluteus stage and in which the forms of both the larva and the adult are more variable than in the Ophiuromorpha. In fact the metamorphosis of echinomorphs from pluteus larvae is remarkably uniform, whether the adults are sea-urchins, heart urchins, or sand dollars. If the larval phase is regarded as a later addition to echinoderm life histories, it may be supposed that this phase was acquired by one ancestor of all modern echinomorphs, and that, although the larval and adult forms have since shown considerable evolutionary divergence, the method of metamorphosis has remained much the same as in this postulated ancestor. A similar larval form, it is suggested, was acquired by one representative of each of several families of ophiuromorphs. Each might have acquired its larval phase at about the same time, or there might have been gaps of millions of years during which the larval form underwent some changes. Even if the genetic package received by each of the ophiuromorphs concerned prescribed the same larval form, however, it would have included no detailed instructions for metamorphosis, and each recipient in each family would have had to devise its own method. Later larval evolution would account for the differences that can be observed today.

Why sea-urchins and brittle-stars should both develop from pluteus larvae forms the subject of the next chapter.

8

Echinoderms: Sea-Urchins and Brittle-Stars

Sea-urchins and brittle-stars very different as adults, very similar as pluteus larvae—Echinoplutei and ophioplutei only larvae with calcareous skeletal rods—Convergent evolution rejected as explanation—Postulated that echinomorph acquired larva from asteromorph then evolved skeletal rods (first pluteus); ophiuromorph acquired pluteus from echinomorph—Some biochemical similarities between adult sea-urchins and brittle-stars attributable to growth of early juveniles within pluteus larvae—Three pairs of species of echinoplutei show marked similarities but adults assigned to different orders—Different types of pluteus larvae within genus *Lytechinus* (Echinomorpha) and within genus *Ophiura* (Ophiuromorpha)—Some brittle-stars have doliolaria larvae—These examples attributed comparatively recent genetic transfers affecting larval form.

Adult sea-urchins, with their rounded, rigid bodies and spacious coeloms, look very different from adult brittle-stars, with their snake-like arms and restricted coeloms, but the majority of species in both classes have a unique form of larva, the pluteus. Its uniqueness lies in its internal skeleton of slender calcareous rods that supports its ciliated arms. Hyman (1955, p. 700), in her volume on *The Echinodermata*, thought it impossible to account for the occurrence of pluteus larvae with similar skeletal rods in both echinomorphs and ophiuromorphs "except on the basis of some community of ancestry," and a similar view was expressed by Jägersten (1972) in his book *Evolution of the Metazoan Life Cycle*. MacBride (1914, p. 511), in his *Text-Book of Embryology*, went so far as to say that the differences between echinoplutei and ophioplutei were "of minor taxonomic importance, and would be such as one would expect to find separating the larvae of two families." He perhaps went rather too far, but his comment does illustrate that some specialists in echinoderms and their larvae have been very impressed by the similarities between echinopluteus and ophiopluteus larvae. Some other specialists, however, have given less weight to larval characters as indicators of evolutionary relationships and have reached a totally different conclusion. Thus Fell (1963) thought that "echinoderm larvae must have followed independent clandestine evolution, in response to temporary planktonic food-gathering phases in the life-history, and do not reflect the relationships of the classes. To infer that [ophiuromorphs] are more closely related to [echinomorphs] than to [asteromorphs], as their larvae would imply, is too preposterous to warrant further serious consideration." (I have substituted my own names for the classes.) Some modern classifications of the phylum have followed Fell in grouping the Ophiuromorpha with the Asteromorpha in the Asterozoa, and the Echinomorpha with the Holothuromorpha in the Echinozoa, although Smith (1984) has taken the Ophiuromorpha out of the Asterozoa and placed them closer to the Echinozoa. Smith's compromise classification is an attempt to give due weight to both adult and larval characters, which is entirely logical if the echinoderms have reached their present state, as adults and larvae, purely by descent with modification. If, however, as I propose, larvae were a relatively late addition to the life histories of echinoderms and spread from class to class, then Fell's views on the classification of the phylum become totally acceptable, even if his suggestion that the larvae have undergone clandestine evolution is questionable.

The term "clandestine evolution" was coined by De Beer (1930), in his book *Embryology and Evolution,* to cover evolutionary change in larvae or juveniles with little or no accompanying change in adult form. In the case in point the larval changes that Fell envisaged must have been toward a similar end product, as he later (1968) implied by attributing them to convergent as well as clandestine evolution. The insertion of a new larval phase into an existing life history, which I am now proposing as the explanation of many anomalies, would provide a larva where there was none before without changing the form of the adult. This may, therefore, be regarded as an extreme form of clandestine evolution, but it is certainly not what either Fell or De Beer had in mind. De Beer did not mention clandestine evolution in relation to pluteus larvae, but he apparently did regard the resemblances between echinoplutei and ophioplutei as convergent, stating that they are "spurious and due to the adaptive needs of flotation," and a number of other authors have expressed similar views. We must, therefore, consider seriously whether the similarities between the larvae of the two groups are more apparent than real and whether they could have arisen as a response to environmental pressures.

Echinoplutei and ophioplutei show many similarities, but there are also some important differences. Fully developed larvae of both groups usually have four pairs of arms (Figs. 5.3, 8.1, 8.2, 9.3), but some families of echinoplutei have more, and the order of development of the arms is different in the two classes. In a developing echinopluteus the first arms to appear are the postorals, followed by the anterolaterals, posterodorsals, and preorals. The postorals usually remain the largest, the posterodorsals are usually the outermost pair, and the preorals often remain small. The larvae of several families have posterolateral lobes, and in the Arbacioida (Fig. 8.1c) and Spatangoida (Fig. 8.1d) these are produced into long arms that make their first appearance after the anterolaterals and before the posterodorsals. Spatangoid larvae also have a pair of anterodorsal arms and a single aboral spike, giving them a total of 13 arms. Ophioplutei never have preoral arms; the posterolaterals develop first and remain the largest, and the posterodorsals develop last; sometimes there is a small aboral prominence but never a well-developed spike. Unidentified species have been described with a small posterior projection or small dorsal and ventral projections near the base of each posterolateral arm, but there are no known ophiopluteus larvae with more than eight arms. Larvae with less than eight arms, or in which some pairs are greatly reduced, are known in both classes (Fig. 8.1f,g). "Epaulettes" are short lateral portions of the ciliated band that separate from the main band and become thickened and arched and provided with very long cilia. They occur in a number of echinoplutei and in at least one ophiopluteus. The supporting skeletal rods arise in the same way in both classes. They start from one or more granules in the mesenchyme, which become triradiate or tetraradiate; these then elongate and may put out further branches. In echinoplutei the rods arise from two centres of calcification on each side of the body, and also from an unpaired anterior centre and sometimes an unpaired posterior centre. In *Clypeaster humilis*, however, before the development of the pluteus is complete, several of the skeletal structures disappear, leaving only those derived from one centre of calcification on each side. The resulting rather simplified skeleton is not unlike that of a typical ophiopluteus. In this class there is always only one centre of calcification on each side, which supplies a branch for each of the arms on that side. A median anterior rod is known only in some unidentified ophioplutei, grouped by Mortensen (1921) under the name *Ophiopluteus costatus,* and there is no unpaired posterior centre of calcification. In many but not all echinoplutei the rods of the two main arms are fenestrated. Fenestrated rods are rare in ophioplutei but do occur in *Ophiura texturata.* The rods may be smooth or bear lateral spines in either class.

Some echinoplutei, but no ophioplutei, have muscles that can be used to vary the angles between the longer arms on the opposing sides. Echinoplutei of the families Cidaridae and Diadematidae are capable of very large arm movements. Cidarid larvae are discussed in Chapter 9 in another context, and one of them is drawn (Fig. 9.3), but now attention is drawn to the larva of a diadematid, *Diadema setosum,* shown in Figure 8.1f. In this echinopluteus the movable postoral arms are very long and also broad and flat, and all the other arms are greatly reduced. The intermittent swimming movements of the larva reminded Mortensen (1931) of the pulsations of a hydromedusa, such as *Obelia.* The ophiopluteus of *Ophiothrix* also has one pair of greatly developed arms (the

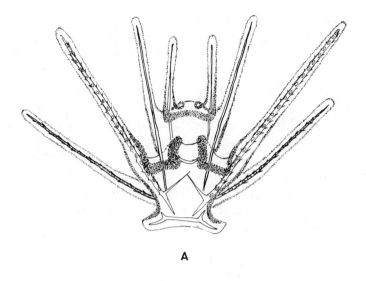

A

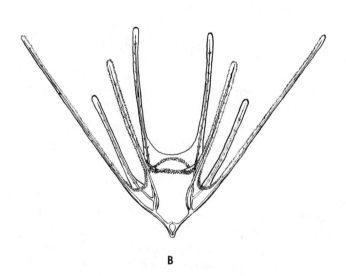

B

FIG. 8.1. Some pluteus larvae: b and g are ophioplutei, the others echinoplutei. (a) *Echinometra lucunter;* (b) *Ophiopluteus compressus;* (c) *Lovenia elongata;* (d) *Temnotrema scillae;* (e) *Arbacia* sp (anterior rods broken); (f) *Diadema setosum;* (g) *Ophiothrix* sp. (Redrawn from Mortensen, 1921, 1931, 1937.)

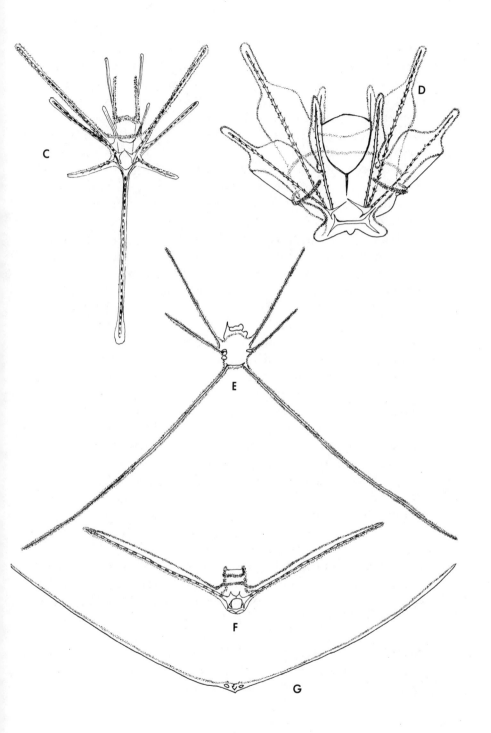

85

posterolaterals) and the others greatly reduced (Fig. 12g), but here the long arms are fixed in position and not flattened, and the larva relies solely on its cilia for propulsion. Fell (1968) cited the larvae of *Diadema* and *Ophiothrix* as examples of convergent evolution, but convergence is restricted to the number of arms. The shapes of the arms and the uses the larvae make of them are quite different. Pluteus larvae of both classes show considerable variation in the number and shape of arms, and the larvae of *Diadema* and *Ophiothrix* provide an example from each class of extreme reduction in the number of functional arms.

This example illustrates that limited convergence between echinopluteus and ophiopluteus larvae can take place, but much more relevant to the present debate is the question of whether echinopluteus and ophiopluteus larvae in general could have reached their broadly similar form as the result of convergence. Certainly we can dismiss the argument that the similarities are a direct result of following a similar planktonic way of life in the same environment. Anyone who has spent a significant amount of time sorting marine plankton cannot fail to be impressed by the amazing diversity of shapes he sees, with forms ranging from spherical to lamelliform and projections varying in number from zero to hundreds, and covering almost every conceivable variation in size, configuration, and orientation. The prime necessities for planktonic larvae are food and some control over their depth, but provided they remain small (and most of them do) their environment seems to exercise little restraint on their shape. Certainly there is no general tendency for planktonic forms to resemble miniature wigwams, easels, or bilaterally symmetrical shuttlecocks. The only ones to have this shape and the only ciliated ones to have endoskeletons of slender calcareous rods are echinopluteus and ophiopluteus larvae.

Not only do planktonic organisms in general show enormous diversity in shape but echinoplutei and ophioplutei also cover a wide range in form. Variations in the numbers of arms has already been mentioned, and these arms can be long or short, slender or broad, tapering or club shaped. Only some of the known variations are shown in Figures 5.3, 7.1, 8.1, 8.2, and 9.3. Shapes and differences in shape can rarely be quantified, but it is my subjective assessment that the range of shapes shown by fully formed echinoplutei is considerably greater than that shown by adult echinomorphs. The diversity in shape of ophioplutei is less than that of echinoplutei but is nevertheless greater than that of adult ophiuromorphs. The facts seem quite consistent with the suggestion that pluteus larvae of both groups have reached their present lack of uniformity by divergence from a common ancestral form but inconsistent with the suggestion that their similarities are the result of convergence.

The bipinnaria larva of a starfish has ciliated lobes corresponding to all the arms of pluteus larvae except the aboral spike of spatangoids. I postulate that after many eons without larvae, an echinomorph acquired bipinnaria-like larvae from an asteromorph, and that one of its descendants evolved the capacity to grow skeletal rods, thus producing the first pluteus larvae. I suggest that this new form of larva spread widely among the echinomorphs and diversified with time, and that an ophiuromorph acquired the genetic recipe for a larva of the pluteus type from an echinomorph. I previously thought (Williamson, 1988a) that there was insufficient evidence to determine whether the pluteus type of larva was first evolved by the echinomorphs or the ophiuromorphs, but the range of variation shown by echinoplutei is certainly greater than that shown by ophioplutei. Although there is only a weak correlation, greater diversity seems more likely to reflect a longer period of evolution.

It has been the theme of this chapter that the similarities between the larvae of sea-urchins and brittle-stars point to a close relationship whereas the morphology of the adults (living and fossil) of the two groups seems to indicate only a distant relationship. The suggested resolution of this anomaly is the acceptance of the ideas that the larvae are closely related but the adults are not, and that one group acquired its larval form from the other quite late in its evolutionary history. If we look at the biochemistry of the adults, however, we find some features that appear to give some support to the groupings suggested by larval but not adult morphology. In general, the phosphagens, sterols, and collagens of adult echinoderms indicate affinities between sea-urchins and brittle-stars and between starfish and sea cucumbers (Kerkut, 1960; Goad et al., 1972; Matsumura et al., 1979). (It will be recalled that the bipinnaria larva of a starfish shows considerable resemblance to the auricularia larva of a sea cucumber.) These biochemical features of adult echinoderms may, however, give misleading clues to the way in which they have evolved. Not all the phospagens, sterols, and collagens of all species investigated support the groupings suggested by the majority, and the sterols at least can be modified by controlling the diet of the adult echinoderm. Environment, therefore, may play a significant role in the development of some of these chemicals. The authors quoted, however, assume that the constitution of the majority of the compounds investigated is determined genetically. Natural selection for such chemicals is likely to have favoured genes coded for substances that can be made from ingredients readily available to the early juvenile stages of each group. As we have seen, these early juveniles develop as quasiparasites within their respective larvae. I suggest, therefore, that the chemical similarities between adult sea-urchins and brittle-stars and between adult starfish and sea cucumbers largely reflect the similar diet and chemical environment enjoyed by their quasiparasitic juveniles within their larvae. This implies that the similarities between echinopluteus and ophiopluteus larvae are not limited to their morphology but also include their chemical composition, and the same applies to the similarities between bipinnaria and auricularia larvae.

So far I have considered the Echinomorpha and Ophiuromorpha as classes, pointing out that the pluteus larvae of the two groups seem to indicate a close relationship while the adults appear to be much more distantly related. There are also a few examples within each of these classes of species that appear to be closely related as larvae but not as adults or closely related as adults but not as larvae. De Beer

(1951) gave some examples, originally listed by von Ubisch, of apparently similar pairs of echinopluteus larvae derived from adults sufficiently dissimilar to be classified in different orders or superorders. He stated that "the structure of the skeleton of the four-armed pluteus is practically identical in *Sphaerechinus granularis* and *Echinocyamus pusillus*" (species from different superorders), that "the four-armed pluteus of *Arbacia lixula* is practically indistinguishable from that of *Strongylocentrotus franciscanus*" (species from different orders), and that "the eight-armed pluteus of *Arbacia punctulata* is very similar to that of *Echinocardium cordatum*" (species from different superorders). In these cases, however, the larval similarities are limited to the pattern of spines and fenestrations of the skeletal rods and do not extend to the overall shape of the larvae. The larva of *Arbacia punctulata*, for instance, does not possess the aboral spike that is a feature of the larva *Echinocardium cordatum* and other heart-urchins, such as *Lovenia* (Fig. 8.1). De Beer thought that the similarities in the skeletons in question may be the result of similar physiological conditions within the larvae and may not reflect close genetic affinity, but experiments in hybridization between different species of sea-urchins have shown that the form of the skeletal rods of the pluteus larvae is generally inherited and may show maternal or paternal features or a mixture of both (Giudice, 1973; Horstadius, 1973). I regard these anomalies listed by De Beer (1951) as cases of clandestine evolution brought about by recent genetic changes that have affected some features of the larvae but have left the adults virtually unchanged. They are of particular interest because, in each case, the unexpected features of a larva are known to occur in another larva that is apparently only distantly related. These cases of echinomorph larvae with incongruous endoskeletons are broadly parallel to that of the spider crab *Dorhynchus,* discussed in Chapter 4, which has an incongruous exoskeleton, and they are all consistent with the suggestion that recent horizontal transfers of genetic material have affected larval but not adult features.

The foregoing examples are of unexpectedly similar features found in the larvae of dissimilar adults, but some of the known examples of dissimilar larvae of similar adults are perhaps even more striking. *Lytechinus variegatus* occurs in the Caribbean and *L. amnesus* and *L. panamensis* are found on the Pacific coast of North America. Mortensen (1921) reared larvae of all three species and found them all quite similar, as one would expect in species of the same genus. The illustrations of a 15-day-old pluteus of *L. variegatus* shown in Figure 8.2a are taken from his drawings. The larva is fully formed, with four pairs of arms as well as posterolateral lobes, and the skeletal rods are not fenestrated and do not fuse together to form a "basket structure." *Lytechinus verrucula- tus*, a sea-urchin with a wide distribution in the Indian and Pacific Oceans, was also reared by Mortensen and its larva described in a later publication (1931). Early development, as far as the gastrula, was rapid, but the arms developed unusually slowly. There were still only two pairs of arms when the 20-day-old larva was drawn, and there were no signs of posterolateral lobes (Fig. 8.2b). Not only was the outward appearance of the larva very different from that of *L. variegatus* and the other species of the genus, but the skeleton was also quite different. Mortensen made detailed drawings of the skeleton when the larva was preserved at 28 days, but there had been only slight development since the larva was drawn 8 days previously. The postoral rods were fenestrated for most of their length, and the elements from both sides were fused posteriorly to form a "basket struc- ture." This is a form of skeleton that is well known in a number of echinopluteus larvae, but it is very different from that of the other species of the genus *Lytechinus*. In fact, the larva of *L. verruculatus* does not seem to belong to the genus *Lytechinus*.

Mortensen (1931) also drew attention to a comparable case of incongruence between adult and larval characters in the Ophiuromorpha. *Ophiura albida* and *O. texturata* are com- mon around the coasts of Scandinavia and the British Isles. Their adults show only minor differences and appear to be two closely related species, but their pluteus larvae are very clearly different. The pluteus of *O. albida* has very broad arms, with no two pairs of the same length, whereas that of *O. texturata* has narrow arms with all except the posterodorsals ending at the same level, and whereas the larval arms of *O. albida* are supported by spiny rods with no fenestrations, the rods of *O. texturata* bear few spines, and those supporting the outermost pair of arms are fenestrated throughout their length (Fig. 8.2c,d). This seems to be the only known case of fenestrated rods in an ophiopluteus larva. On the basis of larval characters, these two species would be placed in different superfamilies rather than in the same genus.

In spite of their obvious differences, the larvae of *Ophiura albida* and *O. texturata* are both clearly plutei, with arms supported by well-developed calcareous skeletons. Brittle-stars of the genus *Ophiura* are quite similar to those of the genus *Ophiolepis* and are usually placed in the same family, the Ophiolepidae. The larva of *Ophiolepis cincta*, however, is clearly a doliolaria, with no arms and no trace of a calcareous skeleton (Fig. 8.2e). It differs from the better known doliolarias of the Holothuromorpha and Crinomorpha in having only three ciliated bands and two of these are not continuous. An ophiuromorph doliolaria with four continuous bands of cilia is the larva of *Ophioderma brevispinum*, a brittle-star of the family Ophiodermatidae. This could easily be mistaken for the larva of a crinomorph.

In any classification based purely on the characters of embryos and larvae, *Lytechinus verruculatus* would have to be removed from the genus *Lytechinus*, *Ophiura albida* and *O. texturata* have to be placed in different superfamilies, both *Ophiolepis* and *Ophioderma* would have to be removed from the class Ophiuromorpha, and "Kirk's ophiuromorph," which as we saw in Chapter 7 develops as a schizocoelous protostome, would have to be removed from the phylum Echinodermata. If, on the other hand, we confine ourselves solely to adult characters, there seems no reason to question that *Lytechinus verruculatus* belongs to the same genus as *L. variegatus*, *L. amnesus*, and *L. panamensis*, that *Ophiura albida*

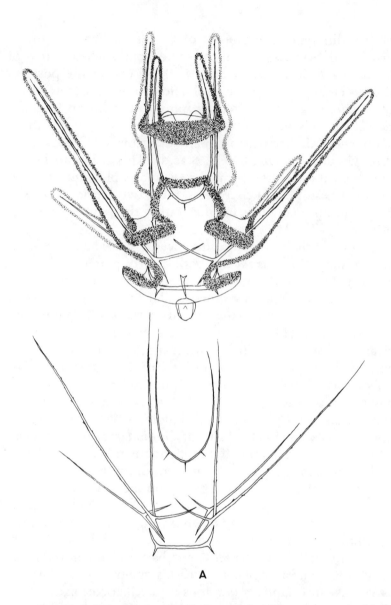

A

FIG. 8.2. Dissimilar echinoderm larvae from adults of the same genus or family. (a) A 15-day-old echinopluteus of *Lytechinus variegatus* and its skeleton; (b) 20-day-old echinopluteus of *Lytechinus verruculatus* and skeleton of a 28-day-old specimen; (c) fully developed ophiopluteus of *Ophiura albida;* (d) fully developed ophiopluteus of *Ophiura texturata;* (e) doliolaria larva of *Ophiolepis cincta.* (a–d redrawn from Mortensen, 1921, 1931; e redrawn Fell, 1968.)

LARVAE AND EVOLUTION

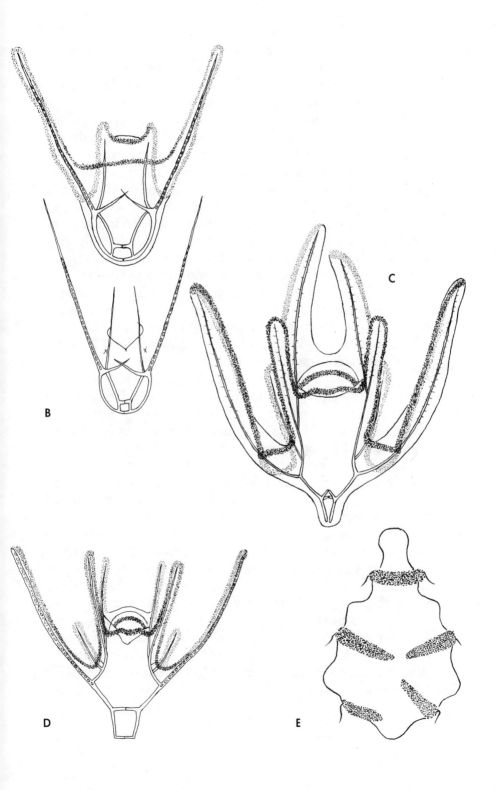

B

C

D

E

and *O. texturata* are members of the same genus, that *Ophiura* and *Ophiolepis* are members of the same family, or that *Ophiolepis, Ophioderma* and Kirk's species are all members of the Ophiuromorpha. All these anomalies between the form of larvae and embryos on the one hand and adults on the other seem to be examples of marked changes in early development that have taken place without any marked change in adult features. Conventional phyletic theory, which assumes that all parts of the life history of a species must have evolved together throughout their phylogeny, can offer, at best, only unconvincing explanations of these anomalies, but they are fully consistent with the idea that packages of genes affecting the form of embryos and larvae can occasionally be transferred from one species to another, which is not necessarily closely related. "Kirk's ophiuromorph" is not only a species with no larva but, it seems, one whose ancestors neither evolved nor acquired one. *Ophiolepis* and *Ophioderma,* with doliolaria instead of pluteus larvae, might have acquired their larvae quite recently, but from crinomorphs rather than from other ophiuromorphs. Alternatively, like dromioid crabs, they may be animals that once had larvae of the "expected" type for their group but later evolved direct development and, still later, acquired a larval form from a distantly related group. *Lytechinus verruculatus* and *Ophiura texturata,* each with larvae of the expected pluteus form but with marked differences from those of their nearest relatives, might also have lost their original larvae and then acquired new ones, but, if so, it is not clear where the new larvae came from. If, however, a species with a larva were to acquire genetic material affecting larval form from another species, a larva with a mixture of characters might result. It seems possible that several of the larvae discussed in this chapter come under this category, including not only *L. verruculatus* and *O. texturata* but also the examples listed by von Ubisch and De Beer of echinopluteus larvae with skeletal rods closely resembling those of species in different orders or superorders. The larva of the spider crab *Dorhynchus,* described in Chapter 4, provides another case of an apparent chimeric larva, with a mixture of features, although, as in the echinoderm examples, the adult shows no hybrid features. The possible occurrence of chimeric hybrid larvae will be discussed further in Chapters 12 and 13.

Apart from "Kirk's ophiuromorph," which has no larva, the foregoing section has drawn attention to anomalies in larval form at lower taxonomic levels, including among species of the same genus. The process that originally gave a bilaterally symmetrical larval form to a radially symmetrical echinoderm probably happened several hundred million years ago. The same sort of process, I suggest, gave larvae to some echinomorphs more than a hundred million years ago and to some ophiuromorphs about a hundred million years ago, although the spread of larvae among the ophiuromorphs is still continuing, and a few seem never to have acquired them. The examples listed here provide evidence that the same process has been at work in recent geological time, introducing new larval forms into existing life histories and modifying existing larvae. It must be presumed to be still at work, producing its effects in parallel to the gradual evolution of species from other species.

9
Echinoderms: Fossil Record

Earliest known echinoderms radially symmetrical—Partial bilateral symmetry
of sea-cucumbers and heart-urchins probably of recent origin—Postulated that
no echinoderms had larvae before Carboniferous—Extinction of many classes
in Triassic but four classes which survived might have had larvae—Brooded
embryos and large genital apertures provide evidence of direct development in
fossils—Existing evidence consistent with present theories but inconclusive

The statement that the echinoderms evolved from bilaterally
symmetrical ancestors occurs in many textbooks of invertebrate zoology, but the only evidence that bilaterally symmetrical adult echinoderms ever existed comes from modern
echinoderm larvae, and the theme of much of this book so
far is that this evidence is open to an entirely different interpretation. Those who claim that bilaterally symmetrical adult
echinoderms must have existed at one time will probably
suggest that such animals were soft bodied and left no fossils
or that the relevant fossils have yet to be found. The earliest
known fossil echinoderms, however, were not bilaterally
symmetrical and gave no hint that they might have been derived from bilateral forms. Some of the earliest known fossils
of the group are shown in Figure 9.1, and only some of them
were pentaradial. The curious *Helicoplacus* (Fig. 9.1b) was
a triradial form with its body spirally twisted, and Paul and
Smith (1984) suggested that all the pentaradial echinoderms
evolved from ancestors such as this. *Stromatocystites* (Fig.
9.1c) shows that pentaradial forms were already present in
the Lower Cambrian, but some Cambrian echinoderms
showed a symmetry intermediate between triradial and pentaradial, and it seems probable that pentaradial echinoderms
were derived from triradial ancestors. *Tribrachidium*, which

occurs in Precambrian strata, showed triradial features, but too little is known of its structure to be sure that it was an echinoderm.

Cambrian strata include some asymmetrical forms, including the cinctans (Fig. 9.1d) and the carpoids (sometimes grouped together as calcichordates or homolozoans), which show resemblances to radial echinoderms. It has been suggested that asymmetrical forms gave rise to both echinoderms and chordates, but this theory seems to run into difficulties over chronology. Jefferies (1979), a proponent of the theory, suggested that "echinoderms and chordates probably separated from each other later than the beginning of the Cambrian, when skeletons began. (This is a rare occasion when it is possible to say: 'later than . . .'),'' yet both tri- and pentaradial echinoderms seem to have been established by the Lower Cambrian (Paul, 1979 and Fig. 9.1), and *Tribrachidium*, which Paul regards as a probable echinoderm, was a member of the late Precambrian fauna. The existence of early asymmetrical fossils does, however, raise the possibility that radially symmetrical echinoderms had asymmetrical ancestors. At present this hypothesis is no more than a possibility, but it does not conflict in any way with the proposal in this book that, for much of their history, echinoderms were radially symmetrical throughout life and had no distinct larval stage. It gives no support to the rival theory that the earliest echinoderms were bilaterally symmetrical throughout life and their larvae have remained so.

The fossil record also gives no support to those who suggest that the partial bilateral symmetry of modern sea cucumbers and heart urchins might be derived from a bilaterally symmetrical remote ancestor. There is good agreement among palaeontologists that all modern echinoderms can be traced back to Cambrian ancestors that were radially symmetrical, and, whereas sea cucumbers and heart urchins are probably not the first to evolve some bilateral features, such features have no long evolutionary history.

Palaeontology not only gives us information on the shapes of extinct echinoderms but also on their numbers. The his-

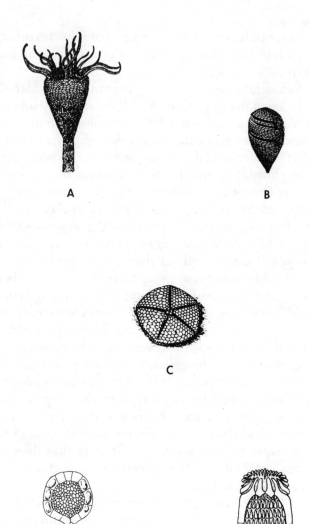

FIG. 9.1. Early echinoderms. (a)*Kinzercystis* (Eocrinomorpha); (b) *Helicoplacus* (Helicoplacomorpha); (c) *Stromatocystites* (Edrioasteromorpha); (d) *Trochocystites* (a cynctan); (e) *Ctenocystis* (Ctenocystomorpha); (f) *Gogia* (Eocrinomorpha); (g) *Macrocystella* (Cystomorpha). (a–c) Lower Cambrian; (d–f) Middle Cambrian; (g) Upper Cambrian. (From Paul, 1979.)

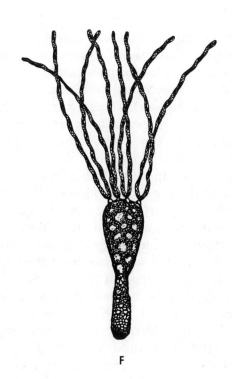

F

G

tory of the phylum is made up of periods of diversification and periods of mass extinction, the two processes sometimes going on simultaneously, with some classes producing increasing numbers of species, genera, and families and other classes were being wiped out (Fig. 9.2). These processes were very conveniently documented by Paul (1979), from whose paper Figure 9.2 is taken. At the class level, there was great radiation in the Cambrian, followed by much slower radiation in the Ordovician, to give an all-time maximum of 19 classes by the end of the Ordovician. (This ignores the Concentricyclomorpha, about whose origins we can only guess.) Only one class, the Blastomorpha, appeared after the end of the Ordovician, and it became extinct in the Permian. There was a steady decline in the number of classes from the early Carboniferous to reach its present level of five (six with the monospecific Concentricyclomorpha) in the late Triassic. The number of genera of echinoderms reached a maximum of about 400 at the end of the Carboniferous, then declined rapidly to a minimum of about 40 by the end of the Triassic. Since then, diversity within the surviving classes has increased rapidly, and the number of genera has almost regained the late Cambrian maximum, but no new classes have evolved. This pattern of a Triassic/Jurassic minimum, followed by a marked increase in the number of genera but not of classes, is not confined to the echinoderms but applies to most metazoans.

Unfortunately, no fossil echinoderm larvae are known, but it is possible to suggest a theoretical time scale for the spread of larvae to different groups of echinoderms consistent with what is known of the periods of diversification and extinction of adult echinoderms. To do this, it is convenient to work both backward and forward from the Triassic minimum. Before this period, it is suggested, only a minority of echinoderms in any class had larvae, and those classes that had become extinct probably never had any representatives with larvae. The rapid and wide dispersal available to animals with a planktonic larval phase is particularly valuable in times of rapid climatic change because it greatly increases

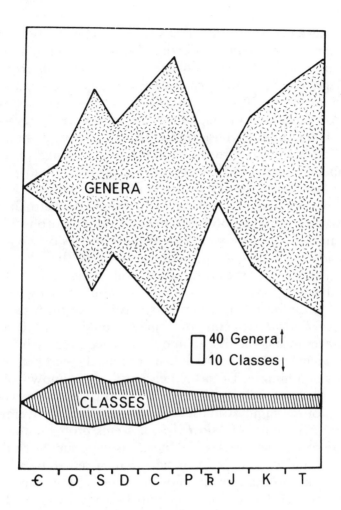

FIG. 9.2. The numbers of genera and classes of fossil echinoderms from the Lower Cambrian to the Upper Tertiary. (From Paul, 1979.)

the chance that some individuals could reach conditions suitable for growth and reproduction. With the exception of the ophiuromorphs and probably the concentricyclomorphs, it is suggested that all the echinoderms to survive the mass extinctions of the Permian and Triassic were species with planktonic larvae, although there may also have been species with larvae among those that failed to survive. A study of the distribution patterns of fossil echinoderms might lead to distinctions between those forms that probably had planktonic larvae and those that probably did not.

Many echinoderms suffered extinction in the Permian and Triassic periods, and probably the only echinomorphs to survive into the Jurassic were one or more species of the genus *Miocidaris* (Clarkson, 1979). It seems likely that their survival was aided by the possession of a planktonic larval phase, and that the considerable range of form of modern echinoplutei is the result of evolutionary divergence from the pluteus of *Miocidaris*. This, of course, would be the case by whatever means the first echinomorphs acquired their larvae and whenever they acquired them, provided it was before the Triassic minimum. Unfortunately, nothing is known of the larva of any fossil species of echinoderm, but, as surviving adults of the family Cidaridae show only limited divergence from fossil adults of *Miocidaris,* it seems probable that the larvae of Jurassic species of *Miocidaris* were similar to those of modern Cidaridae. Such larvae are characterised by a rather simple skeleton supporting movable arms and by numerous, large, ciliated lobes, not supported by rods, as in the larva of *Prionocidaris baculosa* (Fig. 9.3a,b). It was suggested in Chapter 6 that the first echinopluteus developed from a larva resembling an asteromorph bipinnaria, and the cidarid echinopluteus shows the resemblance to a bipinnaria very clearly. In fact, if one can imagine the bipinnaria shown in Figure 9.3c with two pairs of lateral lobes modified into long arms supported by rods, it would bear a striking resemblance to the echinopluteus shown in Figure 9.3a. Comparison of Figure 9.3a and b shows the considerable movement of which the long arms of this echinopluteus are capable, and the lobes of many asteromorph bipinnarias are also movable. (The long processes of the bipinnaria of *Luidia sarsi,* shown in Figure 7.2, seem to work almost like wings.) Many echinoplutei are incapable of moving their arms and have lost this facility, which, I suggest, early pluteus larvae inherited from the bipinnaria-like larvae from which they evolved.

To postulate a time scale for the spread and evolution of echinoderm larvae, it may be suggested that perhaps the development of larval skeletal rods was an innovation that first appeared in ancestors of *Miocidaris* in the late Permian, and that their ancestors in the early Permian had acquired bipinnaria-type larvae from an asteromorph. The first asteromorphs to acquire larvae might have done so in the late Carboniferous, obtaining them from a holothuromorph. The asteromorph bipinnaria is quite similar to the holothuromorph auricularia, from which it is assumed to have developed, whereas the asteromorph brachiolaria, which has organs of attachment and which succeeds the bipinnaria in some but not all asteromorphs, was probably a later addition, allowing the juvenile to crawl away from the remains of a settled larva rather than starting its free existence in mid-water for which it is ill adapted. The auricularia of the Holothuromorpha is the echinoderm larva closest to the tornaria of the Enteropneusta, from which it seems to be derived, and it is assumed to have changed comparatively little since its interphylar transfer, perhaps in the mid-Carboniferous. It is suggested that the doliolaria was developed by one line of holothuromorphs by bringing forward into the late larval phase the translation of genetic factors prescribing radial symmetry that had previously operated only in the juvenile and adult. This enables the metamorphosing juvenile holothuromorph to greater use of larval tissues than its counterparts in any of the other classes of echinoderms. Those crinomorphs with a doliolaria larva might have obtained it from a line of holothuromorphs in which larval development had been abbreviated by the elimination of the auricularia phase, but they devised an entirely new method of metamorphosis, quite different from that employed by the holothuromorphs.

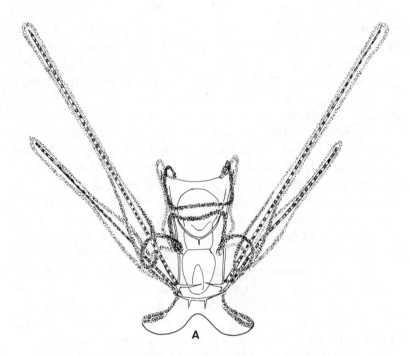

A

FIG. 9.3. Echinopluteus and bipinnaria larvae. (a,b) Echinopluteus of *Priono-cidaris baculosa:* (a) ventral view; (b) lateral view, with the long arms thrown back. (c) Bipinnaria of *Pentaceraster mammillatus,* ventral view. (a,b redrawn from Mortensen, 1938; c redrawn from Fell, 1968.)

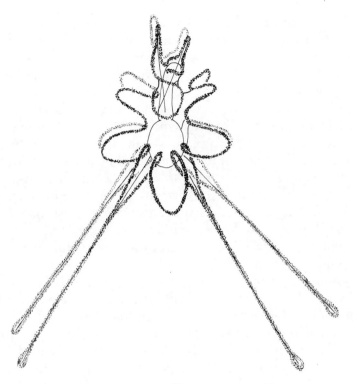

B

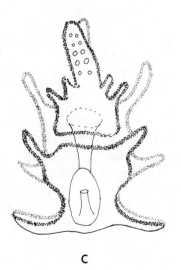

C

The ophiuromorphs, which, as adults, are relatively mobile forms, seem, in general, to have survived drastic climatic changes better than other echinoderms, and it is suggested that the several families that outlived the Triassic minimum did so without the benefit of planktonic larvae. This, it is postulated, was probably the last class of echinoderms to acquire larvae, the first of its members to do so obtaining the genetic prescription for a pluteus from an echinomorph well after the Triassic minimum. From this line derivatives of the original ophiopluteus spread to other ophiuromorphs. The relatively small amount of diversity of ophiopluteus larvae seems consistent with this suggested late acquisition of the pluteus larval form. Some ophiuromorphs, however, do not have pluteus larvae. Some hatch as a nonfeeding doliolaria, for which, I suggest, they probably acquired the genetic prescription from a crinomorph; however, a few, including Kirk's species, seem to have retained the ancestral form of echinoderm development, with no planktonic larva.

The suggested sequence of events proposed in the foregoing paragraphs, although fully compatible with what is known of the diversification and extinction of echinoderms in the past, is nevertheless largely conjectural, but it need not always remain so. There is little prospect of obtaining fossil echinoderm larvae in the foreseeable future, but it is possible, in some cases, to deduce from fossil echinoderms whether they had larvae. The American biologists Jablonski and Lutz (1983) brought together observations from various sources to point out (1) that sexual dimorphism in modern echinomorphs is known only in species in which there is no free-swimming larval stage, and the same may well be true of fossils, (2) that some fossil echinoderms possess a form of brood chamber or marsupium, suggesting that they released benthic juveniles rather than planktonic larvae, and (3) that the diameter of the genital pore reflects the size of the egg and hence the mode of development. Modern echinoderm eggs of less than 0.4 mm in diameter hatch as planktonic larvae that feed on phytoplankton; eggs of about 0.5 mm in diameter may give rise to short-lived planktonic larvae that feed only on internal yolk or, as in the case of Kirk's ophiu-romorph, may hatch as benthic juveniles; eggs of about 1 mm in diameter probably always hatch as benthic juveniles. The genital pores are always wider than the eggs that pass through them, and Jablonski and Lutz quote Müller (1970) as suggesting that the dividing line between planktotrophic and nonplanktotrophic larval development coincides with a genital pore diameter of about 1 mm. Large, fossilised eggs were found in the body of one species of Carboniferous blas-tomorph, providing direct evidence of brooding, and it has been deduced that a late Cretaceous cidarid echinomorph had planktonic larvae.

These observations, mentioned by Jablonski and Lutz, are fully consistent with the history of echinoderms suggested in this book, but are far from sufficient to prove the theories now being advanced. They do, however, illustrate that, in time, sufficient palaeontological evidence may be gathered to test these theories. In interpreting such evidence, it should be borne in mind that brooding can, in some cases, be a sec-ondary evolutionary feature, developed in animals whose ancestors had planktonic larvae. Among the more obvious questions are: Did all the early echinoderms have large gen-ital pores, brooded eggs or other evidence of direct devel-opment, and, if so, when did small genital pores first appear in the various groups of echinoderms?

1 0

Trochophorate Animals: Polychaetes, Echiurans, Sipunculans, Molluscs

Marked differences between Annelida, Echiura, Sipuncula and Mollusca but some members of each phylum have trochophore larvae—Convergent evolution and evolutionary conservatism rejected as explanations of similar larvae—Methods of metamorphosis from trochophore differ from phylum to phylum and from polychaete family to family, but all involve great loss of larval tissue—Postulated that representatives of all four phyla acquired trochophores (or pretrochophores) from another phylum—All trochophores are schizocoelous protostomes but *Viviparus* (a mollusc) is an enterocoelous deuterostome—Postulated that *Viviparus* acquired its embryonic development from another phylum

In earlier chapters attention was drawn to the Echinodermata and the Hemichordata, phyla of animals that, as adults, have virtually nothing in common but that are generally assumed to be offshoots of the same major branch of the evolutionary tree because they have similar larvae. This assumption is, I claim, erroneous, and the evidence from the larvae is open to a quite different interpretation. The Annelida, Echiura, Sipuncula, and Mollusca provide a comparable example. Again the adults of the respective phyla are very different, but at least some of the members of each group hatch as larvae that are so similar that they are given the same name, trochophore. Again it has been argued that the similar larvae reflect the true affinities of the adults, and again I disagree. The trochophore larva will be considered later, but first let us consider the adults of the four phyla that can develop from it.

The Annelida, or annelids, are worms whose bodies are composed of a series of small rings (Latin: anneli). The rings are body segments, and nerve ganglia, excretory organs, reproductive organs, and some blood vessels and muscles are repeated in each segment. The gut runs directly from the anterior mouth to the posterior anus. The coelom also reaches the whole length of the body, but it is typically divided by transverse septa, so that there is one compartment for each segment. Nerves from a dorsal anterior "brain" pass each side of the esophagus to join a ventral longitudinal nerve cord linking the segmental ganglia. The blood systems of the respective segments are linked by both dorsal and ventral vessels, and the dorsal vessel and some of the anterior segmental components may be contractile. Fossil representatives of the phylum go back at least as far as Lower Cambrian strata and probably to the Precambrian. Its modern representatives include the groups Oligochaeta (earthworms), Aeolosomata, Hirudinea (leeches), Branchiobdella, and Polychaeta. Of these, only the Polychaeta (Fig. 10.1a) have larvae, and many of them hatch as trochophores. The polychaetes are nearly all marine worms, and the name of the group refers to the many chaetae (setae or bristles) that project from the parapodia, bilobed structures borne laterally on all or most of the segments. Such worms range in length from a few millimetres to over a metre. The parapodia of swimming forms can act like paddles, but many polychaetes crawl over the substratum or among stones, others burrow in sand or mud, and yet others live in permanent tubes from which they can extend a crown of tentacles. Most of the crawling and swimming forms feed on macroscopic, living prey, but some trap small prey in a net of mucus. The burrowers either pass sand or mud through the gut or "lick" sand grains to get their nutriment, and the tube-living forms filter living plankton or dead organic particles from the seawater.

The Echiura are unsegmented, coelomate, marine worms, each with a proboscis at the anterior end. The phylum takes its name from the genus *Echiurus,* which means "serpent's tail," but the analogy is not obvious. The size and shape of

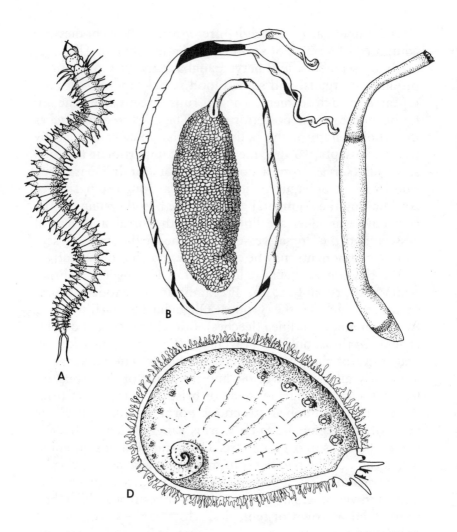

FIG. 10.1. Examples of adult animals that hatch as trochophore larvae. (a) *Nereis* (Annelida, Polychaeta); (b) *Bonellia* (Echiura); (c) *Golfingia* (Sipuncula); (d) *Haliotis* (Mollusca, Gastropoda). (a redrawn from Barnes et al., 1988; b redrawn from Shipley, 1896; c redrawn from Gibbs, 1977; d original.)

the proboscis vary greatly from genus to genus, and it may be anything from 1/35 the length of the body to 20 times as long. In *Bonellia* (Fig. 10.1b) it is long, slender, and bifurcate at the tip, rather like a serpent's tongue, but in *Echiurus* it is short and spoon shaped. *Ikeda taenioides* can have a body of 40 cm and a proboscis of 1.5 but most echiurans are considerably smaller. All members of the phylum have a convoluted gut running through an undivided coelom from the base of the proboscis to the anus at the posterior tip. Many have a closed blood system with dorsal and ventral vessels, but in others the vascular system is "open," i.e., there are no vessels. There is a circumesophageal nerve ring and a ventral nerve cord, but no definite ganglia. Excretion is through many nephridia, which are not arranged in a metameric manner. The dermis bears an anterior pair of hooks and sometimes some posterior setae similar to those of annelids. *Echiurus* has dermal tubercles arranged in rings, giving an impression of segmentation, but such tubercles are not found in other genera, and there are no other signs of segmentation in the adult worms. In some genera the sexes are similar, but in others the males are small, ciliated parasites that live on or in the females. Representatives of the phylum occur in mud, sand, rock crevices, and empty mollusc shells, and they are generally detritus feeders. No definitely identified fossils of the group are known.

The Sipuncula take their name from the genus *Sipunculus*, which means "little pipe." Another generic name, *Golfingia* (Fig. 10.1c), was given by Professor E. Ray Lankester to a sipunculan dredged from St. Andrews, Scotland, apparently alluding to its capture near the home of golf. Like the Echiura, sipunculans are unsegmented, coelomate, marine worms, but they do not have a proboscis, and the anus is not at the posterior end. The mouth, surrounded by short tentacles, lies at the anterior end of an "introvert," which can be retracted into the main trunk. The esophagus, in the introvert, leads to a coiled intestine, in the trunk, and the hind gut terminates in an anus situated dorsally in the anterior part of the trunk. There is no trace of segmentation. The ventral nerve cord is without ganglia and the coelom is undivided. There are no blood vessels. The outer body wall is frequently smooth, but some species have hooks, spines, or tubercles on parts of the epidermis, although these show no sign of metameric arrangement. Different species range from about one to 50 cm in length, and they live in temporary burrows in mud or in empty shells or worm tubes. They feed on small particles picked up by the mouth tentacles. A few Cambrian fossils have been tentatively ascribed to the Sipuncula, but their identification is disputed.

The Mollusca are a huge phylum that includes the snails, slugs, clams, cuttlefish, and octopuses. The Latin word *molluscus* means a soft nut or soft fungus. Although most of its members are marine, it also includes many freshwater and terrestrial representatives. An external protective shell is probably an ancestral feature of the phylum. It has been retained, in a variety of forms, by the majority of modern molluscs but secondarily lost by the minority. A large, muscular, ventral foot is also a typical feature of members of this phylum. It has kept its original function as a crawling organ in many but has become modified for swimming or burrowing in some and reduced or lost in others. The radula, a ribbonlike, rasping organ that can be protracted through the mouth, is another feature of the majority of molluscs, but is lacking in the minority. Most older textbooks refer to the molluscs as coelomate animals because it was believed that they must be descended from forms with a true body cavity. Modern forms, however, have no body cavity other than that provided by the blood sinuses, and there seems to be no good evidence that they ever had coelomate ancestors (Salvini-Plawen, 1985). There is usually an efficient heart, but other blood vessels are found in only one of the eight modern classes recognised here. It has also been suggested that the original molluscs were segmented animals (e.g., Lemche, 1959), but this view is disputed by several authorities on the phylum (e.g., Yonge and Thompson, 1976; Salvini-Plawen, 1985). Some fossil and recent representatives show some metameric or serial repetition of one or more organs, including (in different groups of molluscs) ctenidia (gills), shell plates, gonads, excretory organs, and muscles, but there is no subdivision of the restricted body cavity. This contrasts with the segmentation of annelids, which occurs throughout the phylum, and in which partial or complete subdivision of the coelom is an essential feature (Clark, 1964). I subscribe to the view that several types of metamerism seen in molluscs arose independently of each other, and quite independently of the metamerism of annelids.

Fossil molluscs are known from all periods back to the earliest Cambrian strata. There is no larval phase in the Cephalopoda, which includes the cuttlefish and octopuses, or in the Monoplacophora, a group thought to have been extinct since the Devonian until living examples were obtained from an oceanic trench in 1952, and the form of development of the Chaetodermomorpha (= Caudofoveata) is unknown. Planktonic larvae, but not necessarily trochophores, occur in the development of at least some members of the other five extant classes. These are the Polyplacophora or chitons, the Gastropoda (most with spiral shells but many without), the Bivalvia, the Scaphopoda or tusk shells, and the worm-like Aplacophora (= Solenogastres). True trochophore larvae, however, are known from only a few genera of prosobranch gastropods (e.g., *Haliotis:* Fig. 10.1d), chitons, and bivalves.

A trochophore larva (Figs. 10.2a, 10.3a,d,g, 10.4a) develops from a gastrula similar to that of an echinoderm (Fig. 5.2b), but the blastopore elongates and the larval mouth develops from the anterior lip of the blastopore. A trochophore is, therefore, a protostome, whereas all echinoderm larvae are deuterostomes. It also differs from an echinoderm larva in having an equatorial band of cilia, the prototroch, just in front of the mouth, instead of one encircling the mouth, and there is usually a tuft of cilia at each end of the trochophore. One or more postoral bands of cilia, known as metatrochs, may also be present in later larvae of the Polychaeta and Echiura. The nervous system of a polychaete trochophore, studied in detail by Lacalli (1984), consists of two main parts. An anterior concentration of cells, the apical organ, is connected to the system of nerves that supplies the prototroch, and a largely separate system supplies the pharynx and other oral apparatus and the metatroch. The nervous system of other trochophores is probably similar. Trochophores of the Polychaeta and Mollusca very soon begin to show features of the appropriate type of juvenile, or such features may already be present at hatching, but trochophores of the Echiura and Sipuncula can have a planktonic life of several weeks before the onset of metamorphosis. The majority of sipunculan trochophores are lecithotrophic, feeding only on their internal supply of yolk. The majority of trochophores of the other phyla are planktotrophic, but there are some lecithotrophic examples in each group.

I pointed out in Chapter 8 that there is no general tendency for planktonic organisms to look like pluteus larvae, and I must now point out that there is equally no general tendency for them to look like trochophores. Most biologists who have investigated the trochophores of polychaetes, echiurans, sipunculans, and molluscs have concluded that the larvae of these different phyla develop in the same way and are essentially similar in structure, but a minority have regarded the similarities as more apparent than real and have assumed that convergence has taken place. Few of those who have favoured convergence as the explanation of the similarities have given any evidence for their views. Salvini-Plawen (1980, 1985) listed differences among the larvae of the different groups, although most of these differences are between metatrochophores rather than trochophores. They show that development from the trochophore follows a different course in each group, but they do not show that the basic larva is different. In spite of his emphasis on the differences, Salvini-Plawen did not conclude that the similarities between the larvae concerned are merely superficial. He considered that the term trochophore should be restricted to early larvae of the Polychaeta and Echiura, but he regarded these and the trochophore-like larvae of the Mollusca and Sipuncula as all being derived from a pericalymma or test cell larva, a larval form that will be considered below. I shall continue to use the term trochophore for molluscan and sipunculan larvae of this type, as well as those of polychaetes and echiurans, regarding them all as fundamentally similar larvae, but, even under Salvini-Plawen's view, there must be real phylogenetic affinity between the larvae of the four phyla. It has usually been assumed that these groups must have inherited their similar larvae from a common ancestor, and that the larvae have remained virtually unchanged, whereas the adults have diverged. The evolutionary divergence of the adults is, as we have seen, very great, for although the Mollusca have, apparently, retained an acoelomate condition, the other three groups have developed extensive coeloms, and the Polychaeta have divided the coelom with a series of transverse septa. Also, this evolutionary divergence was already far advanced at the beginning of the Cambrian period. Among the earliest known fossils are examples of several types of segmented annelids and shelled molluscs, so these two phyla at least were on well-separated evolutionary lines over 600 million years ago.

Defenders of the conventional interpretation of the facts concerning these phyla seem to believe that early stages in development are much more conservative in their morphology than later stages, and some claim Darwin as their ally, but it is worth checking what Darwin actually said on this subject. He did point out (1859, p. 419) that the unhatched embryos of distantly related animals are frequently very similar to each other, but he also noted (1859, p. 420) that "the case is different when an animal during any part of its embryonic career is active, and has to provide for itself." (His use of "embryonic" refers not only to the unhatched true embryo but includes the hatched larva.) He goes on to say that "the adaptation of the larva to its conditions of life is just as perfect and as beautiful as in the adult animal." I agree with Darwin that larvae are by no means immune to evolutionary change, and this includes early as well as late larvae. It is certainly true that trochophore larvae cannot possibly show as much variation as adult polychaetes, echiurans, sipunculans, and molluscs, for not only are these larvae all planktonic rather than being adapted to a great variety of habitats, but they have far fewer parts to vary. Nevertheless, there is plenty of evidence that small and simple larvae can show very obvious evolutionary diversity. An instructive parallel is provided by the nauplius larvae of crustaceans, some examples of which were shown in Figure 3.2. These larvae are comparable to trochophores in their size and complexity (or lack of it), but they show such a range of form that the nauplii of most of the major groups of crustaceans can be distinguished at a glance, and specialists in the various groups can identify them to families, genera, and even species. By contrast, it is usually very difficult to assign early trochophores to a phylum. We shall see in the next chapter, however, that there are several types of larvae built on the same general plan as trochophores but clearly differing from them and from each other. Their existence seems quite incompatible with the concept of the immutable trochophore, but the conventional explanation of the similar trochophores of polychaetes, echiurans, sipunculans, and molluscs postulates that this form of larva has been immune to evolutionary change for well over 600 million years.

Some living animals are very similar to fossil forms that existed many millions of years ago. For example, the only living representatives of the Xiphosura, a group of marine arachnids, are two species of *Limulus,* so-called horseshoe crabs. Xiphosurans were relatively common in Lower Palaeozoic strata (Whittington, 1979), and there are comparable examples among the Brachiopoda (Wright, 1979). Darwin (1859, pp. 151–152) mentioned the duck-billed platypus and "anomalous fishes" that "may almost be called living fossils; they have endured to the present day, from having inhabited a confined area, and from having thus been exposed to less severe competition." Each group of "living fossils" is restricted in habitat and and in form, and there appear to be no closely related living groups. By contrast, trochophore larvae occur in the life histories of many hundred genera in four phyla, and, although they are all planktonic, they are found throughout a range of marine habitats. I do not consider trochophores to be living fossils.

The anomaly of adult diversity and larval uniformity in the trochophorate phyla is explicable as another case of the transfer of a larval form from one group to other distantly related ones. The small amount of variation between the trochophores of the different phyla is consistent with the spread of this larval form in fairly recent geological time, leaving it little time in which to diversify.

Although some members of each of four phyla pass through a similar trochophore phase, the methods of metamorphosing from the trochophore are very different in the different groups, and it is difficult to envisage how they could have evolved from one ancestral method purely by species-to-species descent with modification. In the Polychaeta, the details of metamorphosis differ from family to family, but one consistent feature is the enormous loss of larval tissue. "The cerebral ganglion, stomodaeum and midgut are the only larval organs to persist through metamorphosis, undergoing varying degrees of redifferentiation" (Anderson, 1973), which implies that the larval ectoderm makes no contribution to the next phase in development, and "the adult nervous system develops from proliferative centres quite separate from the larval system, and its nerves also follow largely separate paths" (Lacalli, 1984). The segments of the developing juvenile first appear as bands of mesoderm tissue at the posterior end of the larva, and a split in each band enlarges to form the coelom of each segment. The coelom is therefore a schizocoel, and it arises quite differently from the enterocoels of echinoderm larvae. The segmented polychaete may develop by the gradual proliferation of segments at the posterior end of the trochophore to produce a second larva, the polytrochula or nectochaete. In some genera, however, such as *Owenia* (Fig. 10.2), the extrusion of the segmented body of the next phase may be quite explosive. In this genus the late trochophore, known as a mitraria, develops a bunch of stiff setae at the posterior end and the prototroch becomes con-

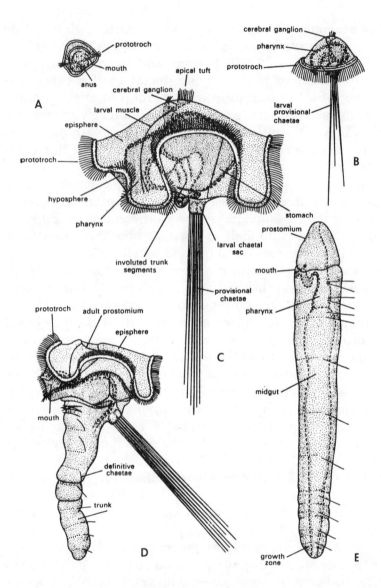

FIG. 10.2. The larval development and metamorphosis of the polychaete *Owenia fusiformis*. (a) Trochophore; (b) young mitraria larva; (c) fully developed mitraria; (d) mitraria 15 seconds after the onset of metamorphosis; (e) first benthic stage, 15–20 minutes after the onset of metamorphosis. (Redrawn from Anderson, 1973, after Wilson, 1932.)

voluted (Fig. 10.2c). The trunk segments of the developing juvenile first form between the mouth and anus of the larva, and then grow around the larval intestine. The trunk rudiment is invaginated in such a way that the first five segments "are turned inside out and drawn back over the succeeding segments much as the top of a stocking can be turned inside out and drawn back over the foot" (Wilson, 1932). At metamorphosis (Fig. 10.2d), the juvenile trunk evaginates in less than 30 seconds and during the next 15 minutes the larval tissues disintegrate and are swallowed by the young worm (Fig. 10.2e). The term "cataclysmic metamorphosis" has been applied to such cases, but the disintegration of the larva is usually less rapid. In some Phyllodocidae about 40 segments of a well-developed, wriggling juvenile may protrude from the anal region of the trochophore, while the prototroch continues to provide pelagic transport. The degree of independence of the larva and the juvenile is reminiscent of that seen in echinoderms. The variable process of metamorphosis in the Polychaeta, in which so much larval tissue is lost that in no case can the larva be said to "develop into" the juvenile, seems to have no satisfactory explanation in terms of traditional neo-Darwinian evolution, but it is fully consistent with the suggestion that members of the group acquired trochophore larvae from some other group comparatively recently.

In the Echiura, metamorphosis from the pelagic trochophore to the settled juvenile involves considerable changes, but the metamorphosis is gradual, not cataclysmic (Dawydoff, 1959; Pilger, 1978). As the larva elongates, the ciliated bands are lost and the mouth moves to near the anterior end, but a preoral lobe persists and grows to form the proboscis. The length of the gut, and particularly the intestine, increases enormously, and larval excretory organs, if present, are replaced by adult nephridia. The adult coelom is derived by expansion of the larval coelom, which appears to be derived from the blastocoel. The coelom shows no trace of segmentation, but several genera show clear but transitory signs of metamerism at the time of metamorphosis. In *Echiurus*

(Fig. 10.3a), *Lissomyema* (= *Thalassema*) and *Bonellia* the paired ventral nerve cords develop a series of ganglia, and at either side of the nerve cords are a series of bands of mesoderm tissue, although these do not correspond to the ganglia. In *Echiurus*, but not the other two genera, each mesodermal band develops its own small schizocoel, very much as in the Polychaeta. In the metamorphosing *Urechis*, 12 annular rings of ectodermal mucous glands encircle the body at intervals from the mouth to the anus, and give a clear impression of segmentation (Gould-Somero, 1975) (Fig. 10.3c), although the internal organs show no trace of metamerism. The different genera of the phylum thus show varying degrees of metamerism, in some cases affecting different organs, but in all genera these apparent signs of segmentation totally disappear as the worm grows.

The writers of earlier textbooks of invertebrate zoology (e.g., Borradaile et al., 1935) generally assumed that the Echiura had evolved from annelids by complete loss of segmentation in adults, although traces of it persist in metamorphosing larvae. If this were the case it would imply that unsegmented echiurans had supplanted their segmented ancestors by natural selection, but, as they are burrowing worms, it is not clear why the unsegmented forms were selected. Indeed, Clark (1964) has argued that a major factor in the success of the annelids is the segmentation of the coelom in that group. There is little agreement among recent authors on the relationship of the Echiura to the Annelida, and none of them offers any explanation of the transient metamerism of metamorphosing echiurans. This phenomenon, however, is explicable if it is assumed that the Echiura, previously without a larval phase, acquired their larvae from one or more annelids. I suggest that, in each case, they acquired the genetic recipe not only for an annelidan trochophore larva but also for the development of a metameric adult with a septate schizocoel. Such segmentation, however, plays no part in the body plan of an adult echiuran, and the only way they could complete their metamorphosis was to suppress the metamerism. It must be assumed that different evolutionary branches of the Echiura were already established, that each acquired its larvae independently, and that each independently devised its own means of suppressing segmentation. Some, it would seem, have completely suppressed metamerism of the ectoderm whereas others have suppressed it in the mesoderm. The segmented mesoderm, when it occurs, may be lost before or immediately after the development of schizocoelic pockets.

A few sipunculans develop directly, without a larval phase, but most hatch as trochophores. Two species have been reported to metamorphose directly from trochophores to benthic juveniles, but a further 14 species are known to pass through a second planktonic larval phase, the pelagosphaera, and one species is known to hatch in this form (Rice, 1975). Some pelagosphaeras are, as their name implies, spherical, but others are elongated to varying degrees. This larva differs from a trochophore in having no preoral prototroch, or only a vestigial one. Most pelagosphaeras rely for propulsion on a well-developed postoral metatroch, but in a minority there are no obvious bands of cilia. The mouth is anterior and the anus dorsal, as in the adult, and most of the adult organs develop during this phase. Lecithotrophic pelagosphaeras settle within a few days, but planktotrophic ones may remain planktonic for several months. One species of pelagosphaera, probably belonging to the genus *Sipunculus,* can reach a diameter of 10 mm, but most are considerably smaller. Metamorphosis from the trochophore, whether resulting in a pelagosphaera or a settled juvenile, always involves the loss of most of the ectoderm (Fig. 10.3c–e). Hyman (1959) mentions nineteenth-century reports of metameric bands of mesoderm with coelomic cavities that appear briefly during the development of *Phascolopsis gouldii,* much as in some echiurans, but more recent work has failed to confirm this transient segmentation (Rice, 1975). Whether or not there is any trace of metamerism in the Sipuncula, I suggest that the pelagosphaera was the original planktonic larva of the phylum, and that a trochophore stage was acquired later, perhaps from a polychaete. The large loss of tissue at meta-

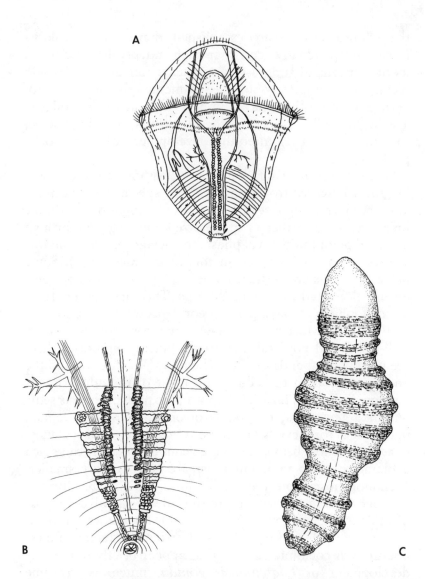

FIG. 10.3. Trochophore larvae and some subsequent developmental stages in the Echiura and Sipuncula. (a–c) Echiura: (a) late trochophore of *Echiurus* in ventral view; (b) enlarged view of posteroventral region of a similar larva; (c) newly metamorphosed *Urechis caupo*. (d–f) Sipuncula; (d) early trochophore of *Golfingia;* (e) late trochophore of *Sipunculus,* shedding outer membrane; (f) pelagosphaera of *Sipunculus*. (a, b redrawn from Sedgwick, 1898, after Hatschek; c redrawn from Gould-Somero, 1975; d–f redrawn from Rice, 1975, d after Gerould, e after Hatschek.)

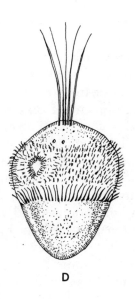

D

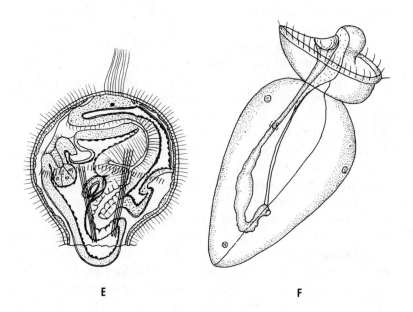

E   F

morphosis from the trochophore seems consistent with such an interpretation.

Most textbooks of invertebrate zoology state or imply that the trochophore is the ancestral form of larva in the Mollusca, but trochophores or trochophore-like larvae are known to occur in the development of only a few genera of prosobranch gastropods, chitons (Polyplacophora), and bivalves. A prosobranch trochophore is shown in Figure 10.4a. Polyplacophoran trochophores are similar, but feed only on internal yolk and lack an anus. It is much more usual for a gastropod or a bivalve to hatch as a veliger, a larva with a shell, and with cilia borne on a lobed velum, an organ specialised for locomotion and feeding. In the late veligers of some gastropods the velar lobes may be greatly extended (e.g., Fig. 10.4c). In *Clione* and some other gymnosomatous pteropods in which the adult has no shell, the early veliger may still have a well-developed shell. The trochophore stage, when it occurs, is always followed by a shelled veliger, although in this case the velum is circular and develops directly from the prototroch (Fig. 10.4b). Jägersten (1972) insisted that a veliger is merely a modified late trochophore, but I cannot share his view. In species that hatch as a trochophore, the ciliated cells of the velum develop directly from those of the prototroch, but all other organs of the trochophore are discarded by the veliger or drastically modified. Practically all the adult organs develop during the veliger stage, but the velum is purely a larval organ and is completely shed at metamorphosis from the veliger to the juvenile.

The veliger is a very common molluscan larva, and many bivalves have a bivalved veliger. Veligers, however, do not occur in all bivalves with planktonic larva, and members of the subclass Protobranchia hatch as a lecithotrophic larva with three bands of cilia and a single external aperture at the posterior end. The juvenile develops within this larva, and the larval ectoderm, often referred to as a test, is completely shed at metamorphosis, much as in the Sipuncula. A larva that undergoes such a metamorphosis is known as a test cell larva or a pericalymma. Some members of the class Aplacophora develop directly, without a larva, but others hatch as a lecithotrophic test cell larva that has a single ciliary band and apical tuft, as in a trochophore, but resembles a protobranch larva in having only one exterior opening, the posteriorly situated "pseudoblastopore" (Fig. 10.4d). The juvenile develops from cells around this aperture, at first internally, then externally, and produces its own ring of cilia, the telotroch. The larva degenerates as the juvenile grows, and the larval test breaks up into rounded cells that are incorporated into the coelom at the anterior end of the juvenile (Fig. 10.4e–i) (Thompson, 1961). A modified form of test cell larva also occurs in the Scaphopoda.

As mentioned above, Salvini-Plawen (1980, 1985) suggested that the ancestral larval type in the Annelida, Echiura, Sipuncula, and Mollusca was not a trochophore or trochophore-like larva but a test cell larva, and that this gave rise to a trochophore or pseudotrochophore in some groups. Certainly test cell larvae occur in several groups of molluscs that are usually regarded as primitive, but if these groups have retained the ancestral type of larva, would they not also be expected to retain the ancestral type of metamorphosis? I would argue that the cataclysmic type of metamorphosis that links test cell larvae and their corresponding juveniles cannot possibly be ancestral, and it is difficult to see how it could have arisen by the accumulation of successive mutations. The relatively few larval cells that contribute by direct descent to the body of the juvenile is a feature of the metamorphosis of several other groups with what I regard as introduced larvae, and I suggest that the test cell larvae of molluscs have also been acquired from another group. Salvini-Plawen's suggestion that the molluscan trochophore evolved from a test cell larva may well be correct, and I would add the suggestion that the line that modified the larva in this way also modified the method of metamorphosis to reach the following stage without cataclysmic disruption. Nevertheless, in spite of this modification, virtually none of the tissues and organs of the molluscan trochophore is retained by the veliger, and, in this respect, metamorphosis in the

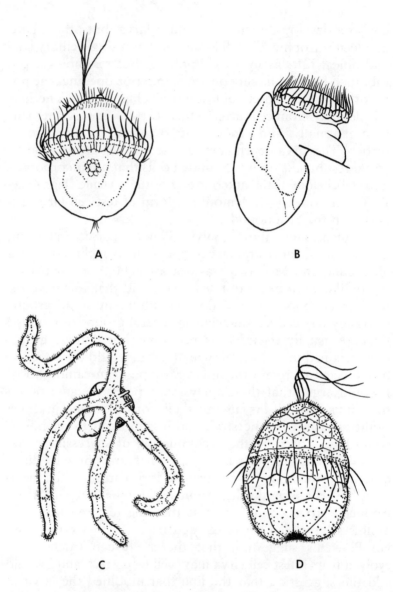

A   B

C   D

FIG. 10.4. Stages in the development of some molluscs. (a, b) Trochophore and veliger of *Acmaea testudinalis* (Prosobranchia); (c) late veliger of a prosobranch gastropod; (d–i) stages from newly hatched larva to juvenile of *Neomenia* (Aplacophora). (Redrawn from Jägersten, 1972; a,b after Kessel, 1964; c after Dawydoff, d–i after Thompson, 1961.)

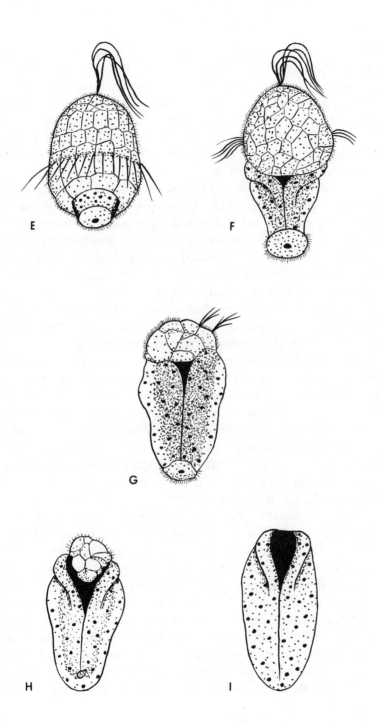

Mollusca is comparable to that in the other trochophorate phyla. I have no new evidence to add to the debate on whether the molluscan trochophore (or pseudotrochophore) evolved from a test cell larva or vice versa, or whether they evolved independently, but I suggest that either or both are larval forms introduced into the Mollusca by horizontal genetic transfer from another group, or that one was introduced and the other evolved from it.

We cannot leave the Mollusca without mentioning the strange case of *Viviparus viviparus,* a freshwater prosobranch gastropod with direct development. At least six biologists have published accounts of the embryology of this species (references in Fernando, 1931), and all are agreed that, in contrast to the vast majority of molluscs, it is a deuterostome, in which the blastopore becomes the anus. Some of the authors have claimed that the mesoderm originates from the ectoderm, but Fernando's detailed and careful study leaves no doubt that it originates from the endoderm. The ventral wall of the archenteron bulges out and extends laterally to form a coelomic sac that separates from the archenteron. The sac continues to grow, then disintegrates into separate cells that multiply to fill the body cavity. *Viviparus,* therefore, starts development as an enterocoelous deuterostome but ends up as a conventional mollusc. It will be recalled that the Echinodermata, which are usually regarded as enterocoelous deuterostomes, include "Kirk's ophiuromorph," which develops as a schizocoelous protostome, and I have suggested that this represents the method of development of ancestral echinoderms. Kirk's species and *Viviparus* are exceptional in their respective phyla, but this is no reason for regarding both as showing ancestral forms of development. Indeed the pattern of development of *Viviparus,* involving a transient enterocoel, seems quite unsuited to a mollusc, and I suggest that the pattern has been recently acquired from an enteropneust, an echinoderm, or a chaetognath. Be that as it may, the existence of two types of early development in both the brittle-stars and the prosobranch gastropods points to a change in the genetic coding for development in each group, whether it has affected the minority or the majority. In the proso-branch gastropods, this change must have taken place after the establishment of the group as a distinct evolutionary line, that is comparatively late in the history of the molluscs, and it must have had little or no effect on the form of the adults. If the method of mouth formation and coelom formation in the embryo can be drastically altered without any appreciable alteration in adult form, it seems a relatively small step to suggest that the form of a larva can be altered or a new larval form introduced by a similar process. A possible mechanism will be discussed in a later chapter, but, in the meantime, the heterogeneous assemblage of phyla each containing some members with trochophore larvae provide further examples consistent with the concept that transfers of genetic material affecting embryonic and larval development can, from time to time, take place between distinct evolutionary lines.

In earlier chapters it was suggested that the echinoderms had no planktonic larvae until they acquired a larval form from another group, thereby gaining a means of dispersal. The same probably applies to the Echiura, which, it is postulated, had no planktonic larvae until they acquired a trochophore phase from the Polychaeta. The other three trochophorate phyla, however, each has a dispersal phase that today follows the trochophore in their respective ontogenies but that probably preceded it in their phylogenies. Thus the segmented polychaete larva (nectochaete or polytrochula), the sipunculan pelagosphaera, and the molluscan veliger were probably each part of the life histories of their respective groups before these life histories included a trochophore stage. It looks, therefore, as if those members of these three groups that acquired a new form of larva would have no immediate advantage over their relations with no trochophore stage, but once they had completed their metamorphoses they would be at no disadvantage. The probable method of spread of such larvae is discussed in Chapter 12, and, if my theories are correct, the spread would continue unless there were active selection against the introduced larval form. In time, those animals with the introduced trochophore phase might have gained some advantage over their rivals, for this simpler and smaller first larva would have permitted the evolution of smaller and more numerous eggs, whereas the original larva (now second in ontogeny) could grow at the expense of the late trochophore, and enter the plankton at a rather later stage of development.

I I
Near-Trochophorate Animals: Flatworms, Nemertines, Bryozoans

Planarian flatworms, nemertine worms and bryozoans have larvae resembling
trochophores—Polyclad larvae metamorphose by differential growth—
Juveniles of nemertines and bryozoans develop inside larvae, with much loss of
larval tissue—Suggested that nemertines, bryozoans and all groups with
trochophore larvae (Ch. 10) acquired their larval forms, directly or indirectly,
from polyclads

The occurrence of trochophore larvae in four phyla considered in the previous chapter has frequently been quoted as
evidence of common ancestry, although I claim that the evidence has been wrongly interpreted. Much the same applies
to larvae resembling trochophores and the phyla in which
they occur (the near-trochophorate phyla). I shall confine
myself to larvae whose affinities to trochophores are widely
accepted, thus limiting myself to three phyla, but I would
point out that there are a number of other larval types that
various authors have regarded as being related to trochophores.

In the first phylum, the Platyhelminthes or flatworms, a
number of rather similar larvae occur in members of the Turbellaria Polycladida, the adults of which will be considered
below. In these larvae much of the body may be ciliated, but
the main natatory cilia are in a preoral band that extends
into 8, or occasionally 10, lobes in forms known as Müller's
larva (Fig. 11.1), and four lobes in Götte's larva and some
similar forms. The affinity of polyclad larvae to trocho-

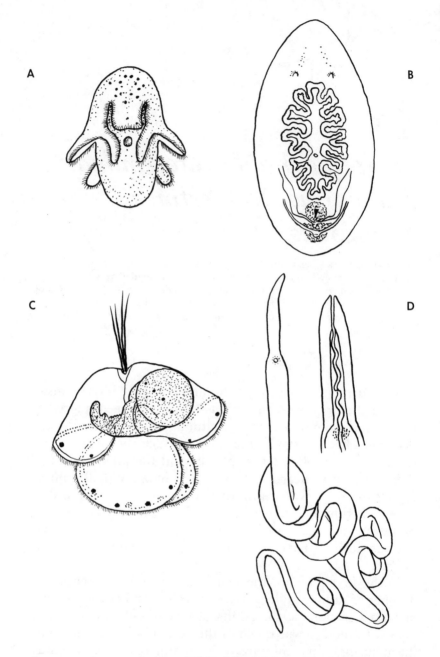

FIG. 11.1. Larvae and adults of the Turbellaria (phylum Platyhelminthes), Nemertea, and Bryozoa. (a, b) Turbellaria: (a) Müller's larva; (b) adult polyclad turbellarian. (c, d) Nemertea: (c) pilidium larva containing developing juvenile (stippled); (d) adult *Procephalothrix*, with enlarged view of head showing part of proboscis. (e–g) Bryozoa: (e) larva of *Alcyonidium*; (f) cyphanautes larva of *Membranipora*; (g) adult zooid of *Electra*. (a, f redrawn from Jägersten, 1972, a after J. Müller; b–d redrawn from Hyman, 1951, c after Verrill; e redrawn from Brien, 1960, after Barrois; g redrawn from Hyman, 1959, after Marcus.)

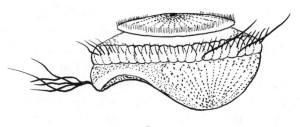

E

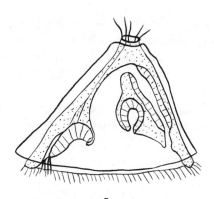

F

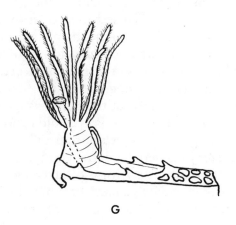

G

phores has long been accepted and is confirmed by the general similarity of the nervous system to that of a larval polychaete, although the system is simpler in Müller's larva (Lacalli, 1984). The existence of larvae that are related to trochophores but are not trochophores and the occurrence of a range of larval form within the Polycladida can be interpreted as demonstrating that such larvae can show evolutionary divergence. The concept of trochophores that have been inherited virtually unchanged by polychaetes and molluscs from a Precambrian common ancestor should be abandoned.

The first verse of Walter Garstang's poem *Mülleria and the Ctenophore,* published posthumously in 1966 but probably written in 1922 or before, is extremely relevant to the views I am now putting forward. Here it is:

Johannes Müller's larva is the primal Trochophore
That shows how early worms grew up from fry in days of yore:
No drastic metamorphosis!—each youngster keeps her skin:
Her larval frills are not thrown off, but eaten from within.

Now Garstang did not, as far as I am aware, contemplate the transfer of larval forms between phyla, but his suggestion that the trochophore had its origin in the life history of a polyclad flatworm is very close to my own thinking on the subject. There is, as he says, no drastic metamorphosis in the polyclads. The larva really does "develop into" the juvenile, and the larval nervous system is not discarded but merely developed and modified. I suggest that the first trochophore in the four phyla considered in the previous chapter was acquired from a member of the Polycladida, not by descent from species to species in Precambrian times but by an interphylar transfer of genetic material comparatively recently. The first to acquire this new form of larva might have been either a member of the Polychaeta or a member of the Mollusca. In either case, it is envisaged that the polyclad larva became modified into a trochophore in the descendants of the individual that first acquired it, before spreading to other members of the same phylum and to other phyla. Whether the original trochophore evolved within the Polychaeta or within the Mollusca, the Echiura and the Sipuncula probably acquired their trochophores from members of the Polychaeta.

The phylum Platyhelminthes contains many parasites, but the members of the class Turbellaria are mostly free living, in the sea, fresh waters, and on land, and trochophore-like larvae occur in one order of this class. The name Turbellaria is derived from a Latin word, *turbellae,* meaning disturbance, but the worms glide extremely smoothly over the substratum or the surface film. The name, in fact, refers to the very local disturbance, visible only under a microscope, made by the cilia that cover the surface of the body. Like all platyhelminths, they are bilaterally symmetrical, and the body consists of three tissue layers, with mesenchyme cells or parenchyma filling the space between the organs. They thus have no coelom, and they are unsegmented. A turbellarian has a mouth at the end of a pharynx that protrudes from the ventral side of the body, and there is usually a gut but never an anus. The larvae in which we are interested all occur in marine and estuarine species of the Polycladida, an order of turbellarians that takes its name from the many-lobed gut of the adult flatworms (Fig. 11.1b). They are typically bottom dwellers, but some can swim, and some are confined to floating *Sargassum.* They are always dorsoventrally flattened and usually ovoid, but they can range from almost circular to at least 10 times as long as wide, with lengths from a few millimetres to several centimetres. Most species have many eyes. The nervous system consists of an anterior "brain," a pair of ventral nerve cords, without ganglia, and a network of smaller nerves. The muscles are not striated and run in circular, longitudinal, diagonal, and transverse fibres. Polyclads are hermaphrodites. They pair for mating, and the eggs are fertilized in the oviducts or in the vagina immediately before laying. Eggs may hatch as a Müller's larva, as a Götte's larva or similar form, or as a juvenile worm. Larval and direct development may occur in different species of the same genus, and in one species with direct development a modified Müller's larva has been found in the egg capsule. It seems probable that development by way of a planktonic larva was the original method in the Polycladida and perhaps in all the Turbellaria, but direct development has evolved independently in many lines.

The pilidium larva of nemertine worms (Fig. 11.1c) has been widely regarded as showing affinities with Müller's larva and the trochophore (e.g., Hyman, 1951; Jägersten, 1972). It is another larva first discovered by Johannes Müller and was named by him after the Greek for a close fitting cap. All pilidium larvae are more or less cap shaped or helmet shaped, but they cover a wide range in the form of the part that would cover the skull and of the "ear-flaps" .hat hang below. The main natatory cilia are borne around the margins of the lobes ("ear-flaps"), and there is an apical tuft of longer cilia. Not only the general form of the larva but also the detailed structure of the ciliary bands and their associated nerves are quite similar in the pilidium and in Müller's larva (Lacalli, 1982), but the method of metamorphosis from the larva to the juvenile could hardly be more different. A juvenile polyclad develops from a Müller's larva by a smooth process of differential growth, but a juvenile nemertine develops round the stomach of a pilidium like a parasite within the larva, with a totally new orientation (Fig. 11.1c) (Gontcharoff, 1961). Seven or eight discs, representing only a small proportion of the larval ectoderm, invaginate and cut off from the rest of the ectoderm and migrate inward. They then grow, spread, flatten out, and finally fuse together to enclose the larval gut and form the epidermis of the developing juvenile. Although the larval gut is incorporated into the juvenile, the cells undergo considerable redifferentiation, and these, together with the derivatives of the ectodermal discs, are the only larval cells that contribute to the juvenile. As metamorphosis approaches, the juvenile shows independent movements within the larva and eventually emerges from it.

The pilidium, now without its quasiparasite, goes on swimming for some time, demonstrating an independence of larva and juvenile comparable with that found in the echinoderms. In *Lineus* and some other genera there is no free larval stage, but a modified pilidium, known as Desor's larva, develops within the egg membrane. The juvenile develops within Desor's larva in much the same way as it would in a free-living pilidium.

Some groups of animals have, apparently, managed to incorporate a larval form from another group into their life histories without a cataclysmic metamorphosis to link the acquired larva to the next stage in development. Nevertheless, when a cataclysmic metamorphosis does occur I regard this as a very strong indication that a "foreign" larval stage has been incorporated, and I should certainly regard the pilidium as a larva acquired by the nemertines from another group, probably the polyclad turbellarians.

Nemertine worms make up the phylum variously known as the Nemertea, Nemertina, Nemertinea, Nemertini, or Rhynchocoela (Hyman, 1951; Gontcharoff, 1961). All the names of the phylum except the last are derived from the name of one of its genera, *Nemertes,* a name given originally to a Mediterranean sea nymph. The name Rhynchocoela might be thought to imply that the worms have a true body cavity or coelom, but they have not. The rhynchocoel is a cavity lined with mesoderm, like a conventional coelom, but, unlike a conventional coelom, it does not enclose the digestive tract, gonads, and other organs. It is longitudinal and dorsal in position, and the only organ that it encloses is an eversible proboscis (when it is retracted). This proboscis, which is separate from the gut, and its sheath form the main diagnostic characters that separate the group from other bilaterally symmetrical worms lacking a true body cavity. The body may be cylindrical or partly flattened and is covered by a ciliated epidermis. The mouth is situated ventrally near the anterior end, and there is a posterior anus. Gut pouches, excretory organs, and gonads are usually serially repeated, but there is no coelom other than the rhynchocoel. The rhynchocoel is not segmented. There is a closed, mesodermal vascular system, and a nervous system, consisting of a brain and usually three longitudinal cords. The sexes are usually separate, and fertilization is usually external. Only a minority of genera have a free larva (pilidium) or an egg larva (Desor's larva). Nemertines have great powers of regeneration, and some species regularly reproduce asexually by fragmenting into a number of pieces, each of which will grow into a complete worm. Most nemertines are benthic marine predators that capture prey, often larger than themselves, with the proboscis, and then engulf it whole. Other species, however, are pelagic, and a few live in fresh waters or on land. The group includes the world's longest worms, some exceeding 30 m in length, although the majority are less than 20 cm.

There is considerable variation in larval form among the bryozoans, which may have shelled or unshelled larvae (Hyman, 1951; Brien, 1960; Jägersten, 1972). Among the unshelled forms is the larva of *Alcyonidium* (Fig. 11.1e), which resembles a dorsoventrally flattened trochophore, with a large apical organ and a prominent equatorial ring of cilia corresponding to the prototroch. *Bugula* has an almost spherical, unshelled larva. The name cyphonautes is given to the shelled larvae found in several genera. A well-known example is *Membranipora* (Fig. 11.1f), in which the larva is laterally compressed, triangular in lateral view, the ciliary girdle is ventral, and a bivalved shell develops soon after hatching. *Flustrellidra* has a nonfeeding larva, with no mouth or digestive tract; the shell is rounded apically and the length is more than twice the height. Feeding cyphonautes larvae may remain in the plankton for at least 2 months, but the forms without shells and the nonfeeding larvae have much shorter planktonic lives. There is no larval coelom other than the blastocoel. After settling, all bryozoan larvae embark on a cataclysmic metamorphosis, contracting under a layer of ectoderm into a rounded or oval mass in which all the tissues undergo histolysis. Part of the ectoderm invaginates, then closes off to form a vesicle, and the inner part of this chamber then constricts off a second vesicle, which remains in contact with the first.

The cells surrounding the outer vesicle give rise to the adult lophophore (see below) and pharynx, and those surrounding the inner vesicle produce the remainder of the gut. Nerves, muscles, and an extensive body cavity develop round the gut, but the lining of the coelom has a composite origin, and the cavity itself forms neither a schizocoel nor an enterocoel (Zimmer and Woollacott, 1977). Bryozoans are usually regarded as protostomes (e.g., Fig. 3.1), because the larval mouth, when there is one, develops from the blastopore. The adult mouth, however, is not derived from the larval mouth, and it is my contention that bryozoan larvae were originally acquired from another group. Under this view, there is no basis for classifying adult bryozoans as protostomes. After the completion of the first zooid of a new colony, other members are produced by asexual reproduction.

Adult Bryozoa, or moss animalcules, form encrusting or branched colonies, each of up to a million individuals, on rocks, shells, seaweeds, and a variety of other substrata. They occur in marine, brackish, and fresh waters. The zooids secrete an exoskeleton, into which they can retract and, in many cases, close the orifice with an operculum. A typical zooid (Fig. 11.1g) has a circular or crescentic crown of ciliated tentacles, the lophophore, carrying food particles to the mouth, which opens into a recurved digestive tract ending in an anus situated not far from the mouth but outside the lophophore. The position of the anus gives the phylum its alternative name of Ectoprocta and distinguishes it from the similar group, the Endoprocta, in which the anus is within the ring of tentacles. Bryozoans have a well-developed coelom; one section occurs in the lophophore and extends into the tentacles and another section surrounds the alimentary canal and other organs in the main body. The zooids within a colony can be of several types, with some specialised for feeding and others for grasping, cleaning, or reproduction. The Bryozoa are divided into three classes: the Phylactolaemata, the Stenolaemata, and the

Gymnolaemata. The freshwater Phylactolaemayta have a small prosoma that overhangs the mouth, making them the only trimerous forms in the phylum. The three regions of the body correspond to those of the pterobranch hemichordates (Chapter 6, Fig. 6.1b). The marine Stenolaemata and Gymnolaemata lack a prosoma. They have different muscle systems and different methods of reproduction. Only the Gymnolaemata include species with planktonic larvae.

Lophophores are also found in several other phyla: the Phoronida, the Brachiopoda, the Endoprocta, and the Hemichordata. Hemichordates of the class Enteropneusta, whose possible relationship to the echinoderms was discussed in Chapter 6, have a proboscis instead of a lophophore, but a lophophore is well developed in the other class of this phylum, the Pterobranchia. The Bryozoa are often grouped with these other lophophorate phyla on the basis of their having similar adult characters, but in terms of larval characters they seem to be much more closely allied to the trochophorates or near-trochophorates than the lophophorates. This presents another anomaly that is difficult to explain in terms of conventional evolutionary theory but that is perfectly consistent with the theory that larval forms can occasionally be transferred from one group to another. Conventional theory also has produced no adequate explanation of the origin of the cataclysmic metamorphosis that links the larvae of bryozoans and several other groups with their corresponding juvenile stages, but, as I have suggested before, such drastic measures may sometimes be necessary to develop a juvenile from a larva introduced from another group.

Although I consider that there is strong evidence that the cyphonautes and other bryozoan larvae have evolved from a larval form acquired from another phylum, it is not at all clear which other phylum. Although I postulate that the ultimate origin of all trochophores and larvae showing affinities with trochophores is to be found in the polyclad turbellarians, bryozoan larvae show considerable modification from polyclad larvae. I suggest that this modification probably took place by Darwinian evolution within another group that had acquired the larval form from a polyclad. A bryozoan then acquired this modified larva, now with a bivalved shell, from this group by horizontal genetic transfer. There is much more variation among bryozoan larvae than among true trochophores, suggesting that the former have been evolving longer and were therefore acquired earlier. This, then, seems to rule out the Polychaeta, Echiura, Sipuncula, and Mollusca as the source of bryozoan larvae. We can also rule out the Mollusca Bivalvia on the form of the shell, which not only differs in shape from that of a cyphonautes but has hinged valves, whereas a cyphonautes does not. If it were not for the bivalved shell, it could be postulated that genetic transfer had taken place between a turbellarian flatworm and a bryozoan and that modern bryozoan larvae had evolved from the larva so acquired, but neither turbellarians nor adult bryozoans have bivalved shells. The Brachiopoda are bivalved lophophorates, and the separation of the two classes of this phylum, the Articulata and the Inarticulata, is based on the presence or absence of a hinge between the two shell valves. Some members of the Inarticulata hatch as an unshelled larva that soon secretes the two shell valves, but this larva swims with the cilia of the developing lophophore, and there is a smooth transition from the larva to the settled juvenile. In the Bryozoa, on the other hand, the lophophore, like all other adult organs, is a new structure formed at metamorphosis.

It should be realised that although the theory of horizontal larval transfer has so far failed to provide an explanation for the occurrence of bivalved shells on cyphonautes larvae, these shells are just as difficult to account for in terms of conventional evolutionary theory. Jägersten (1972) has suggested that the remote ancestors of the Bryozoa were bivalved as adults, that in some of these the shell came to develop precociously, in the larva, and that modern cyphonautes larvae have inherited their shells from these ancestral forms. There is, however, no other evidence for the past existence of this hypothetical ancestral group. It is, of course, possible to postulate the previous existence of another hypothetical group whose members had bivalved shells and archaetrochophore larvae, and to suggest that it was a genetic transfer from this group that gave the bryozoans their original larva. I feel, however, that the invocation of hypothetical extinct groups does not lead to very satisfactory explanations, whether these are in terms of conventional evolutionary theory or the new concepts now being proposed.

I readily admit that my theory of horizontal larval transfer does not provide ready answers to all problems concerning larvae, but it does suggest an explanation of apparent larval affinity and adult disparity and also of the occurrence of two distinct types of metamorphosis. The three groups considered in this chapter provide further illustrations of these points. Larval affinities include a preoral ciliated band and similarities of the nervous system. Adult disparities cover the range from acoelomate planarians to coelomate bryozoans, whereas the rhynchocoel of nemertines has some of the characters of a coelom, giving this group a somewhat intermediate position. The metamorphosis of planarians, by differential growth, is in sharp contrast to the cataclysmic processes in both nemertines and bryozoans, in which the juvenile develops from only a small part of the larva and with a totally different orientation.

III
SOLUTIONS

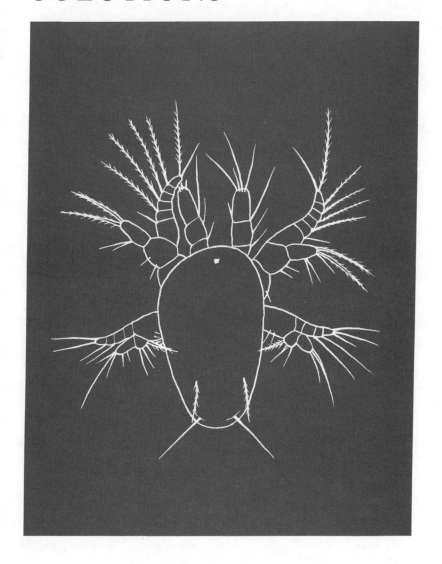

I 2
Horizontal Genetic Transfer

Distinction between larvae evolved within group and larvae acquired from
another—Phases in development which can be transferred between groups—
Transfers between individuals more likely than mass transfers—Hybridization,
usually between forms with external fertilization, most likely method of transfer
—Expected effects of crossing one species with larva and one without—
Expected effects of crossing two species with larvae

The preceding chapters of this book have been largely filled
with examples of larvae that seem out of place in the life
histories of the species or groups in which they occur, but
in most cases the incongruous larva resembles the larva of
some distantly related group. I contend that these similar larvae in distantly related groups are not the result of chance,
convergent evolution, or inheritance from a remote common
ancestor. I believe they have resulted from transfers of genetic material between species on different branches of the
phylogenetic tree, and in the present chapter I wish to discuss
four questions relating to these transfers. First, how can we
distinguish between a larval form that has been introduced
from another group by genetic transfer and one that has
evolved with its corresponding juvenile and adult phases
throughout its phylogenetic history? Second, when there is
evidence of transfer between groups, which phases in development are capable of being transferred? Third, although
whole classes or even phyla can have incongruous larvae, is
this to be regarded as resulting from a mass "infection" of
the group with a foreign larval form, or is it more likely that
the original transfer was between individuals? And fourth,
is there a conceivable mechanism whereby the genetic factors
governing one or more phases in development could have

been transferred from one phylogenetic lineage to another? It must be obvious that the transfer of a larval form from one group to another does not take place often (critics would no doubt add, "or at all"). The form of a larva is inherited, just as is the form of an adult, and occasional minor variations in larval colour or form are inherited, just as are variations in adult colour or form. Larvae are, therefore, subject to descent with genetic variation, just as adults are. Evolution apparently affects larvae no less than adults, and, from considerations such as these, it may be argued that it is just as valid to deduce evolutionary ancestries and relationships from larvae as it is from adults. I contend that the ancestries and relationships of larval forms may be deduced in much the same way as the ancestries and relationships of adult forms, but, when a larval form has been transferred from some other group, the larval and adult phylogenies before the transfer will be different. Very frequently larval and adult phylogenies are quite reconcilable, although there are often some discrepancies resulting from larval and adult characters having evolved at different rates. Such disparities may affect the largely subjective groupings of species into genera, genera into families, and so on, but do not affect the decision as to which forms should be grouped together, regardless of the taxonomic label that is placed on the group. In some cases, however, phylogenies deduced from larval characters are difficult or almost impossible to reconcile with those deduced from adult characters. Such apparent inconsistencies between adult and larval phylogenies provide the first indication that a larval transfer from another evolutionary line may have taken place somewhere in the history of the group concerned.

Ideally, the next indication should be the identification of the original source of the larval form, but although this source is sometimes obvious, it is not always so. The larvae of dromioid crabs, for instance, show very marked resemblances to those of paguroid hermit crabs (Chapter 4), and the auricularia and bipinnaria larvae of echinoderms show marked resemblances to the tornaria larvae of enteropneust hemichordates (Chapter 6). In these two cases the transfers seem to have been from the second group to the first, or, more precisely, from one or more individuals on or near the evolutionary lineage leading to the second group to one or more ancestors of the modern representatives of the first group. On the other hand, it is not, at present, possible to suggest a source for bryozoan larvae, although such larvae fulfil the first proposed test by indicating phylogenetic relationships that are quite irreconcilable with those suggested by adult bryozoans. Difficulties in tracing the source of an acquired larval form could be attributed to great evolutionary divergence in larval characters between the donor group and the recipient group since the transfer, or to the extinction of the donor group.

Further indications as to whether a larval form might have been acquired from another group may come from considering the type of metamorphosis that links the larva to the next phase in development. To say, however, that larval transfer is always correlated with cataclysmic metamorphosis would be a gross oversimplification. I believe that modern dromioid crabs, which are brachyuran decapod crustaceans, are descended from an ancestor that acquired its larva from an anomuran decapod, but the metamorphosis that transforms the last zoeal stage into a megalopa seems much the same as in any other brachyuran or anomuran decapod. In this case, however, the differences between the postulated donor and recipient groups are relatively small, whether we compare zoeas or megalopas. As the two groups concerned are both decapod crustaceans, there are only limited differences in structure between their larvae, so the problems of metamorphosing from an anomuran zoea must be much the same whether the next phase is an anomuran or a brachyuran megalopa. If, as I claim, an anomuran zoea was transferred into the life history of a brachyuran crab, there was no need for a fundamental change in the method of metamorphosis. If, however, the acquired larval phase came from another phylum in either the recent or remote past, the ontogeny, even in modern forms of the recipient group, is likely to give some indication of this dramatic event in phylogeny. Development from the larva is likely to involve profound changes, whether they are sudden or prolonged. Although less extreme transformations may be the result of larval transfers, I certainly interpret the more extreme forms of metamorphosis as indications that a foreign larval phase has been introduced into the life history. This applies particularly when the juvenile develops as a quasiparasite within the larva or when there are changes in symmetry or in orientation during development.

Examples of both sudden and prolonged metamorphosis are found in the polychaete worms, and both are consistent with the introduction of the trochophore larval form from another phylum. Metamorphosis occurs within minutes in *Owenia* (Fig. 10.2), and the epithet "cataclysmic" seems most apt. In other polychaetes, however, such as the Spionidae, metamorphosis from the unsegmented trochophore to the multisegmented juvenile may extend over several weeks; although it is much less dramatic, it still involves massive loss of larval tissue and organs, including virtually all the nervous system.

I regard the loss, as opposed to the modification, of the larval nervous system at metamorphosis, and the simultaneous occurrence of the larval and adult systems at this time, as extremely important indicators that the larva had its origin in a very distantly related group. Dawkins (1986) devotes much of his book *The Blind Watchmaker* to explaining how complex organs can evolve gradually from simple rudiments, with natural selection favouring some mutations at the expense of others. Nature can, however, select variations only in functional organs. A new type of nervous system could not have evolved gradually while the old system continued to function. The occurrence of two nervous systems that overlap in the ontogeny of an animal is best explained by horizontal genetic transfer introducing a new larval phase, with its own nervous system, into the life history of an established adult. I can conceive that, in some groups that undergo a major metamorphosis during development, histolysis and histogenesis might have evolved in a series of short steps, and this seems to have happened in many holometabolous insects. Even in a lepidopteran pupa, in which there is very extensive breakdown of many of the tissues of the caterpillar and remodelling of cells to form the tissues of the butterfly or moth that will emerge, the main (dorsal) blood vessel and the central nervous system are relatively unaffected and pursue an uninterrupted course of differentiation. Such a metamorphosis does not, in my opinion, necessarily mean that the larval form had its origin in another group, although the possibility that caterpillars had their origin in another arthropod group has been tentatively suggested (Chapter 7). When, however, the adult nervous system develops independently of the larval system, I regard this as a prime indication that the larva and the adult had their origins in different phyla, even when there is comparatively little loss of larval tissue, as in the Echiura.

Next, I wish to consider the phases in development that, if my theory is correct, can be transferred from one group to another. The examples of incongruous larvae that I seek to explain as resulting from horizontal genetic transfer include zoeas, tornaria-like larvae, and trochophore-like larvae, so the phenomenon is certainly not restricted to one type of larva. Some of the examples I have given, however, go further and show that not only larvae are involved but embryos as well. It was pointed out in Chapter 5 that all echinoderms with larvae develop as enterocoelous deuterostomes, and this also applies to the majority of forms with direct development. An entirely different form of embryonic development is, however, shown by a minority of brittle-stars, namely Kirk's ophiuromorph (Chapter 7) and members of the Ophiomyxidae and Gorgonocephalidae (Fell, 1941, 1968). These have no trace of larvae and develop as schizocoelous protostomes. There are, therefore, two forms of embryonic development in the echinoderms. The body cavity of all known coelomate metazoans develops either by enterocoely or by schizocoely; deuterostomy is usually associated with enterocoely and protostomy with schizocoely. If the original echinoderms developed either by enterocoely and deuterostomy or by schizocoely and protostomy, some of the modern echinoderms (either the majority or the minority) must have changed. I suggest that this change was the result of the same introduction of genetic material that gave the phylum tornaria-like larvae; in other words, they acquired the genetic recipe not only for a larval form but also for all the steps in the development of the larva from a fertilized egg.

Another example relevant to the present topic is the prosobranch gastropod *Viviparus* (Chapter 10). Nearly all molluscs whose development has been investigated are protostomes, whether they have larvae or not. This snail, however, is out of step with the rest of the phylum in that the embryonic mouth is a deuterostome, and the embryo even develops a short-lived enterocoel. There is no larva. As with the Echinodermata, there has been a change in the method of embryonic development affecting either the majority or the minority of the phylum. In this case, as explained in Chapter 10, I think that it is probably the minority that has undergone the change, but a change there must have been. I regard this as another example resulting from the transfer of genetic material from another evolutionary line, and another example that has affected only development within the egg and not subsequent stages. Introduced genetic material, I believe, produces new forms of development in the target lineage, starting from the fertilized egg and extending a variable distance into subsequent phases of development.

The next question to consider is whether the postulated transfers of genetic material to new groups would have initially affected all or most members of a population, a species, or larger taxonomic group, or whether the original transfer would have been to one individual only. I know of no observation that leads to a direct answer to this question, but it seems unnecessary to postulate transfers of genetic packages affecting more than two individuals (the donor and the recipient) at a time. Thus, as suggested in Chapter 4, all modern dromioid crabs with anomuran-like larvae may be the descendants of a single female. Of course, it can be argued that an event that happened once could possibly have happened more than once, and I do not completely rule out this possibility. On the other hand, a rare event involving two individuals must be rated much more probable than a series of comparable events involving two populations or all the members of two species. Once an individual has acquired a new larval form and has survived through metamorphosis to reproduction, all its descendants will have that same larval form (or one subsequently modified by Darwinian evolution). The spread of the aberrant larval form within the same population and to other populations of the same species is to be expected, provided it confers no selective disadvantage to survival of the recipients.

It was suggested in Chapter 6 that probably only one ancestral echinoderm originally received the genetic prescription for a tornaria-like larva from some ancestor of the modern enteropneusts. Again, the possibility of such an event happening more than once cannot be completely ruled out, but one successful transfer of genetic material would have been sufficient to account for the known distribution of larval forms. Following this event, it is envisaged that there must have been horizontal transfers of genetic recipes for further evolved larvae to individuals of other species of echinoderms, some probably in the same class and others certainly in different classes of the phylum. Again it cannot be ruled out that these transfers took place more than once, but I believe that once would have been sufficient. Similarly, for the other groups with incongruous larvae, I suggest that the horizontal genetic transfers were, in each case, between individuals, and it is unnecessary to postulate that any transfer between two particular species took place more than once. There must, however, have been subsequent horizontal transfers of further evolved larvae between individuals within several of the recipient groups. I have, for example, suggested that the trochophore larval form was acquired comparatively recently by members of several phyla, and also that, after its original acquisition by a polychaete, it was transferred to a member of each of several families of the same group (Chapter 10). "Comparatively recently" refers, of course, to the geological time scale, and the introduction of trochophores to several phyla could have begun any time in the past hundred million years.

The final question to be considered in this chapter is the system of horizontal genetic transfer. If my hypothesis is correct, the mechanism that has produced so many larval misfits must be capable of transferring a genetic package with the programs for both embryonic and larval development between individuals in different species. In many cases, the species are sufficiently distinct to merit classification in different phyla. We know that the features of animals, whether adults or larvae, are largely governed by codes written in chemical sequences strung out along deoxyribonucleic acid (DNA) molecules, mostly on the chromosomes of cell nuclei. These codes are translated from DNA to ribonucleic acid (RNA) molecules within the nucleus, and the RNAs move through the nuclear membrane into the cytoplasm to initiate a series of events that lead to the synthesis of proteins. We also know that, at least in echinoderms and molluscs, many of the RNA sequences found in the cytoplasm of cells during embryonic and larval development are different from those found in the cells of juveniles and adults (Wold et al., 1978; Verdonk and Biggelaar, 1983).

The genetic material in the fertilized egg of any animal with one or more larval phases must carry the codes for the embryo, the first larva, any subsequent larval phases, and, finally, the juvenile and adult, and there must also be a system that ensures that these codes are used in the right order. Actually, the embryo and the first larva are probably covered by the same set of genetic instructions, the embryo merely being the first step in the production of the larva (equivalent to "first catch your hare" in the apocryphal recipe for hare soup). The juvenile and adult are also probably covered by the same genetic instructions, different from those specifying the embryo and larva, with only minor modifications during development. Geneticists may prefer to say that, apart from minor changes, the same set of RNAs is to be found in the cytoplasm of cells throughout the development of the embryo and first larva, and, at the other end of development, another set of RNAs is in operation during the development of the juvenile and adult. The "priority mechanism" will ensure that the genes expressed as the embryo and first larva have priority over over those expressed as any subsequent larval phases, and genes expressed as the final larva will have priority over those relating to the juvenile and adult. This mechanism, I assume, would operate whether the parents were members of the same species or of different species, and cross-fertilizations are particularly relevant. I wish to propose hybridization as the method of horizontal genetic transfer responsible for the cases of incongruous embryos and larvae considered in previous chapters.

Consider what would happen if an egg of a species with direct development (species D, for direct) were to be fertilized by the sperm of another species with a larval phase in its life history (species L, for larvae). The nucleus of the heterozygote would then contain the genetic instructions for all phases in the life histories of both species, but the priority mechanism associated with the production of the larva would be expected to come into effect. This would lead to the development of the embryo of species L that would hatch as the larva of the same species, and development would continue under the influence of the same set of RNAs until a fully formed larva was produced. The hybrid zygote from which, it is suggested, such a larva could develop would have only a haploid set of chromosomes for each species, but a haploid set is apparently all that is needed. It is known, for example, that if an unfertilized sea-urchin egg can be induced to divide parthenogenetically it will go on dividing to produce an apparently normal pluteus larva followed by an apparently normal adult, and I have recently confirmed that the eggs of the cushion star *Asterina gibbosa* will develop readily without fertilization to produce healthy larvae and juveniles. It has been suggested that, in such cases, the cells do not remain haploid (Giudice, 1973), but it has not been demonstrated when, or even whether, they duplicate their sets of chromosomes. This provides a possible explanation of the first part of the larval transfer, that is, how a larva of species L might hatch from an egg of species D and develop until it is ready for metamorphosis.

Metamorphosis would certainly pose problems, but let us consider the possibilities. The cells of the larva would contain the genetic instructions to produce the juveniles and adults of both species L and species D. If the full recipe for two very different species were to operate simultaneously, the result would be every bit as disastrous as trying to mix two very different culinary recipes, but two factors could lead to the development of one animal or the other rather than a mixture of both. First, there are probably genetic factors that, in some cases, give the genes specifying one body form precedence over those specifying the other, in much the same way as a larva generally has precedence over an adult. To achieve the result that I postulate, the development of species D would take precedence over species L, and perhaps a new life history will result only when the genes specifying the body form of the maternal adult take precedence over those specifying the paternal form. If the paternal form took precedence and the hybrid animal continued to develop as species L after metamorphosis, no new life history would result, although one may speculate on the products of backcrossing such hybrids with the maternal species. One may also speculate on the effect of the cytoplasmic contents of the cells of the hybrid, all of which would be derived from species D.

The maternal cytoplasm and nuclear membrane seem likely to provide a second mechanism that would tend to favour the development of one adult form rather than the other, although, in this case, it would always favour the development of the maternal form rather than the paternal. There is widespread agreement among geneticists that cytoplasmic substances play a regulatory role by signalling the expression of specific genes at the proper time (e.g., Rothwell, 1983), and the selective function of the nuclear membrane may be inferred from the observation that RNAs that are found in both the nucleus and cytoplasm of embryos may continue to be produced in the nucleus of adult cells, but most of them are then absent from the cytoplasm (Wold et al., 1978). Let us suppose that the adult body forms of species L and D, the parents of our hypothetical hybrid, are such that there is no factor in their nuclear genes that gives one marked precedence over the other. In the unspecialized cells of the hybrid larva, I suggest that the nuclear membrane and cytoplasm, both derived from species D, would be likely to favour the expression of the genes prescribing the juvenile and adult form of species D rather than species L. It is relevant to point out here that in cases of cataclysmic metamorphosis the juvenile frequently develops from relatively few unspecialized larval cells.

Metamorphosis of a hybrid from an L-type larva to a D-type juvenile could, therefore, result from a natural priority of the body form of type D, or from cytoplasmic factors favouring maternal-type development, or from a combination of both these factors. The result would be an animal that matured as an apparent example of species D but that had a larva resembling that of species L in its life history. Further animals with the same life history would result not only from fertilizations between the gametes of two such LD hybrids but also from crosses between an LD hybrid and a "normal" member of species D, with direct development.

The foregoing scenario could explain how, after hundreds of millions of years without larvae, a female echinoderm with direct development (a species D) produced eggs that were fertilized by the sperm of a species of enteropneust (a species L), so producing the first echinoderms with a larval phase. It was suggested in Chapter 7 that juvenile echinoderms develop from the cells surrounding one or more of the coelomic sacs of the larva because of the configuration of these cells. It is also probably of great relevance that such cells are not modified to carry out any specialized function in the larva, for, as pointed out in the preceding paragraph, only the unspecialized cells of such a larva would retain the maternal form of nuclear membrane and cytoplasm that would be likely to favour the expression of maternal genes at metamorphosis. Metamorphosis in nemertine worms and bryozoans provides other examples of the development of the juvenile from relatively few, unspecialized larval cells.

Egg membranes are known to have properties that favour homosperm fertilization and impede cross-fertilization, but it has long been known that, in some cases at least, such barriers to cross-fertilization can be overcome in the laboratory by comparatively minor changes in pH or salinity (e.g., references in Gray, 1931). This being so, one can envisage that, from time to time, barriers to heterosperm fertilization may be overcome in nature. For example, the chances of successful cross-fertilization are known to be enhanced by exposing the eggs to the action of certain viruses that affect the properties of the egg membrane (Rothwell, 1983), and such viruses certainly occur in nature. Even if the obstacles to cross-fertilization were overcome, further and probably more formidable obstacles would lie ahead at the time of metamorphosis from a larva of one phylum to a juvenile of another. We have seen in earlier chapters, however, the lengths to which some animals will go to achieve a successful metamorphosis, and, in several cases, the postulated change of phylum provides the best explanation of why they go to such lengths.

When an egg of an animal with a larval phase in its development is fertilized by a sperm of another animal with a different larva, this second larva would be added to the life history of the first species only if the genetic instructions for the paternal larva were to take precedence over those for the maternal larva. An example of what I regard as an introduced larva taking precedence over an existing larva is provided by the Sipuncula. Long after the pelagosphaera larva had evolved in a neo-Darwinian manner, the postulated fertilization of a sipunculan egg by a polychaete sperm introduced a trochophore phase that then developed before the pelagosphaera, and it continues to do so in the life histories of the descendants of this hybrid. The introduced larval form is also assumed to have taken precedence over the preexisting larval form in the Polychaeta and Mollusca.

An introduced larval form need not necessarily require a new form of embryonic development, but sometimes it does. It was suggested in earlier chapters that deuterostomy and enterocoely have replaced protostomy and schizocoely several times in the echinoderms and in one known case in the molluscs (in the gastropod species *Viviparus viviparus*). It was also suggested that deuterostomy and enterocoely represent the original form of development in the hemichordates, and hence in the lophophorates, if they form a monophyletic group. This implies that the bryozoans, which are lophophorates, not only acquired a larva from another group but also replaced deuterostomy with protostomy in their early development. The early bryozoans, without larvae, were probably enterocoelous, but bryozoan larvae have no coelom, and the adult coelom is a new structure of composite origin that arises during metamorphosis and cannot be classified as either a schizocoel or an enterocoel. It seems, then, that either method of mouth formation may be replaced by the other, but in all known cases in which the method of coelom formation seems to have changed, the change has been from schizocoely to enterocoely.

Each postulated cross-fertilization leading to the introduction of a new larval phase into an existing life history must depend for its success on a mechanism that promotes the expression of maternal genes and suppresses the expression of paternal genes after the development of the first larva, but not until the first larva is well developed. I know of no experimental cross-fertilization that has clearly shown the change from a paternal larval phase to maternal later phases in development, but experiments described in the next chapter show that a cross-fertilized egg can hatch as a paternal larva. This means that, in this case at least, there is no suppression of the expression of paternal genes in the early stages of development, and the explanations of all the cases of incongruous larvae in this book depend on such lack of suppression.

So far in this chapter we have considered cross-fertilization as a means of introducing a larval phase into the life history of an animal that previously developed directly, or of introducing a new first larva that develops before a larva already established as part of the life history. These cases depend on the introduced paternal larva (and the embryonic development leading to it) taking precedence over existing phases in the maternal life history. What will happen, however, in a cross between two species, both of which have zoea larvae or both of which have pluteus larvae and the priority rating of the paternal and maternal larvae is similar? In such cases one might expect the first larva to show a mixture of paternal and maternal features, but when this larva is well developed the postulated mechanism that favours the expression of maternal rather than paternal genes could come into operation to give an adult indistinguishable from the maternal species. This, it is suggested, is just what has happened in the phylogenies of the spider crab *Dorhynchus* (see Chapter 4), of the brittle-star *Ophiura texturata* (see Chapter 8) and of the several species of sea-urchins also mentioned in Chapter 8, all of which appear to have hybrid larvae but whose adults show no trace of mixed ancestry. These cases all relate to apparent hybridization between pairs of species from different superfamilies, orders, or superorders, and it must be presumed that some minimum degree of genetic difference is necessary before the suppression of paternal genes takes place in later development. It is known that if two closely related species of sea-urchins are crossed they can show a mixture of paternal and maternal characters not only as larvae but also as adults (e.g., Shearer et al., 1914).

In this chapter I have concentrated on those types of hybridization that seem likely to lead to an adult developing from a new form of larva or a new form of embryo. The possible results of other forms of cross-fertilization will be considered in Chapter 14.

External fertilization between shed gametes is the normal method of sexual reproduction in most of the groups under consideration, and cross-fertilization would be most likely to take place between eggs and spermatozoa shed into the sea. Mating takes place in the decapod crustaceans, but fertilization is still external. The male sheds sperm directly over the eggs in the Anomura. In the Brachyura, however, mating and egg-laying are separate events; the female receives one or more spermatophores at the time of mating and sperm is released from these at the time of egg-laying. The postulated cross-fertilizations that produced the hybrid ancestors of the Dromioidea and of *Dorhynchus* probably resulted from cross-matings. The polyclad turbellarians are the only group we are concerned with in which fertilization is internal, but we have considered them only as a probable source of a larval form and hence as a source of sperm in a postulated cross-fertilization. It is not suggested that the polyclads received their larvae from any other group.

This book lists many cases of incongruous larvae that, it is hypothesized, have resulted from transfers of genetic material between distantly related species, and the view has been put forward in the present chapter that these transfers could have resulted from cross-fertilizations. There are a number of ways in which these hypotheses can be tested, and these form the theme of the next chapter.

I3
Toward Confirmation of the Hypotheses

Distinction between theories (1) that larval transfers have taken place, (2) that transfers have resulted from cross-fertilizations—Suggested tests of first theory from fossils, electrophoresis of enzymes, DNA annealing and immunotaxonomy—Suggested tests of second theory from chromosomes, nuclear DNA, ribosomal RNA and experimental cross-fertilizations—Results of two experimental cross-fertilizations, one in literature, one in progress

This book is about two main hypotheses, and, although they are related, it is important to distinguish between them. The first is the concept that embryonic and larval forms have occasionally been transferred between distinct evolutionary lineages, and the second is the suggestion that such transfers have resulted from cross-fertilizations. Distinction should be made between the series of postulated historical events, which is the subject of the first hypothesis, and the mechanism that, it is suggested, was responsible for these events and forms the subject of the second. As I have pointed out elsewhere (Williamson, 1988a), it is generally accepted that dinosaurs are extinct (a postulated historical event) although there is no general acceptance of the mechanism that brought about their extinction. (Birds probably evolved from dinosaurian ancestors, but these ancestors are as extinct as the rest of the dinosaurs.) I am not trying to draw comparisons between the strength of the evidence for larval transfer and that for extinct dinosaurs. I merely use the extinction of dinosaurs as a familiar biological example in which cause and effect are seldom confused. I insist that we should also avoid confusing cause and effect in seeking to explain anomalous larvae and life histories.

Obviously, if it can be demonstrated that embryonic and larval forms can be transferred between distantly related animals by cross-fertilization, it establishes that such events are possible, although it does not prove conclusively that this mechanism has played an important part in shaping life histories. If some or all of my assumptions on the regulation of development in hybrids were shown to be incorrect, this would not prove that hybridization had played no part in shaping life histories and phylogenies. If the cross-fertilization theory were totally discredited, it would not necessarily show that transfers of embryonic and larval form could not take place, for there might be other methods of making the implied transpositions of genetic material. A historical hypothesis can never be completely proved, but it can be disproved by observations and deductions from observations that are inconsistent with it, and it can be strengthened by making predictions from it that are later shown to be true. The present state of both hypotheses now under consideration is that they are consistent with all observations and deductions from observations that I know of, and I submit that they provide the best explanation, and in several cases the only known explanation, of a great many anomalies in larval and embryonic form.

There are a few tests of the larval transfer theory that are independent of the cross-fertilization hypothesis. These include the prediction, included in Chapter 9, that, when a sufficient number of fossil species have been investigated, it will be found that all early echinoderms had comparatively large eggs, consistent with direct development, and that smaller eggs, which probably hatched as planktonic, planktotrophic larvae, did not make their first appearance until about the middle of the fossil history of the phylum. Such investigations should show the subsequent spread of planktonic larvae within the phylum over hundreds of millions of years, although some extant brittle-stars have not yet been affected. Fossil forms with direct development would include not only those that retained the eggs within the body of the parent until they hatched but also others, such as Kirk's ophiuromorph, whose eggs develop directly outside the body of the parent. They would, of course, also include examples of the secondary evolution of direct development in some descendants of species with planktonic larvae. There are good prospects that measurements of genital apertures will provide a basis for distinguishing between those with direct and indirect development. It would, of course, be helpful to know the adult form of Kirk's ophiuromorph, a living species that appears to have retained the ancestral method of development of the echinoderms; but most, if not all, Ophiomyxidae and Gorgonocephalidae probably develop in a comparable manner. If the female genital apertures of these species are significantly larger than those of brittle-stars with pluteus larvae, then the examination of genital apertures of fossil echinoderms is likely to yield crucial evidence.

Several of the techniques of chemotaxonomy and immunotaxonomy should also show the results of gene transfer, by whatever method the genetic material has been transferred. Chromatography, immunology, and electrophoresis may each be used to investigate different groups of chemicals in the tissues of different animals and provide assessments of the relatedness of the animals, and it should be possible to obtain assessments of the relatedness of larvae of different groups and compare them with assessments based on the corresponding adults. In most cases it will probably be necessary to use different chemicals for assessing larvae and for assessing adults, and in many cases it may not be possible to get a direct assessment of the relatedness of larvae to adults of the same species. Nevertheless, it should be possible to get four different assessments of the relatedness of groups based on (1) early planktonic larvae, (2) late planktonic larvae, (3) juveniles, and (4) adults. It may be predicted that, in general, assessments (1) and (2) will be in general agreement, as will (3) and (4), but all four will agree only in cases in which there has been no horizontal genetic transfer in the ancestry of the groups.

Chemical characters are sometimes subject to convergent evolution, just as morphological characters are, and apparent examples are provided by the phosphagens, sterols, and collagens of echinoderms discussed in Chapter 8. If, however, a sufficiently broad spectrum of chemicals is investigated the occasional cases of convergence should become apparent. The enzymes currently widely studied by gel electrophoresis are primary products of gene activity and are probably practically immune to convergent evolution. They should be of the greatest value in investigating anomalies at lower taxonomic levels. Studies over a range of enzyme loci can give quantitative measures of genetic similarity or difference, permitting one to say, for example, that one pair of animals appears to belong to the same species, while another pair appears to belong to different families. Some existing electrophoresis techniques are effective with very small quantities of tissue and could be used with single larvae, but hundreds of larvae of the same species can often be obtained by laboratory rearing and could be used in methods requiring larger amounts of material. DNA annealing and the study of immunological responses are more appropriate for estimating the relatedness of animals at higher taxonomic levels. Any of these methods should give results from early and late larvae of species of different groups that could be compared with those from juveniles and adults of the same species.

Most of the biological groups mentioned in this book as probable recipients of genetic input from another group would form suitable subjects for study by one or more methods of chemotaxonomy. It can be predicted, for example, that the larvae of dromioid crabs would show much closer affinities to larvae of the Paguroidea (Anomura) than to any larvae of the Brachyura, whereas adult dromioids would show closer affinity to adult brachyurans than to adult anomurans. Similarly it can be predicted that echinoderms and enteropneusts would show fairly close affinities as larvae but not as adults, and the same would apply to planarians, nemertines, bryozoans, annelids, echiurans, sipunculans, and molluscs. In the case of annelids, sipunculans, and molluscs, however, a distinction should be made between the trochophore larva, which I regard as having been transferred from another group, and the second larvae, which I should expect to show similar affinities to those of the adults. It may also be predicted that assessments of the relatedness of the various classes of echinoderms based on larvae would be quite different from those based on adults. Enzyme electrophoresis may also be predicted to provide results that confirm, for example, that the adult sea-urchins of the species *Lytechinus variegatus, L. amnesus, L. panamensis,* and *L. verruculatus* are correctly classified as members of the same genus, that the larvae of the first three species are also congeneric, but the larvae of *L. verruculatus* seem to belong to another family.

The theory that the transfers and modifications of larval form that I have postulated have resulted from cross-fertilizations can be tested in a number of ways. The hypothesis implies that any group of animals that has acquired a larval form from another group must have in its ancestry a hybrid between the two groups, and the same would apply to any animals whose larvae show a mixture of characters of two groups. The original hybrid would have had a complete set of chromosomes from each of its parents, and these chromosomes would probably have remained distinct over a considerable number of generations. They would not, however, have remained distinct indefinitely, for it is known that, over a great many generations, the size, shape, and number of chromosomes can change, and this would tend to mask their origins. It should, however, still be possible to identify the hybrid sets of chromosomes in an animal with a hybrid in its recent ancestry. Thus it would be worth comparing the chromosomes of *Dorhynchus thomsoni,* whose larvae were described in Chapter 4, with those of other inachid spider crabs and with homoloid crabs to see whether *Dorhynchus* has any chromosomes not present in other inachids and, if so, whether these give any indication of a homoloid origin. Similarly, we may ask whether the chromosomes of *Arbacia lixula* are similar to those of *A. punctulata* and, if they differ, whether any of the chromosomes of the former species show affinities with some of those of *Strongylocentrotus* and any of those of the latter species with some of those of *Echino-*

cardium. Also it would be worth comparing the chromosomes of *Lytechinus verruculatus* with those of other species of *Lytechinus*, those of *Ophiura albida* with *O. texturata*, those of *Ophiolepis cincta* with other Ophiolepidae, and those of any brittle-stars that develop as enterocoelous deuterostomes (the majority) with those that develop as schizocoelous protostomes (the minority). (These echinomorphs and ophiuromorphs with anomalous larvae were discussed in Chapter 8.) Similar studies may, of course, be made on the chromosomes of any of the species whose anomalous larvae have been regarded as indicating a hybrid in their ancestry, comparing them with members of the group from which the larval form appears to have been acquired and with other members of their own group that do not have this larval form in their life history. This, however, may not be an effective method of obtaining evidence of a very distant hybrid ancestor.

Instead of, or in addition to, investigating the numbers and shapes of chromosomes, studies might be made of the total amount of nuclear DNA in comparable cells of different species. If my hypothesis is correct, each echinoderm with larvae will have the functional nuclear DNA not only to prescribe an adult echinoderm but also to prescribe a larva derived from a tornaria, and it will also have nonfunctional nuclear DNA representing the remains of a disused prescription for an adult enteropneust. This will apply to all direct descendants of the original enteropneust × echinoderm cross, but most echinoderms with larvae will also have other hybrids in their ancestry, and these will also have in their nuclei the remains of disused prescriptions for at least one other adult echinoderm. On the other hand, the functional DNA of Kirk's ophiuromorph and species of Ophiomyxidae and Gorgonocephalidae should be only enough to prescribe an adult echinoderm, and, consequently, considerably less than in their relatives with larvae. These brittle-stars would not be expected to have any nonfunctional DNA inherited from ancestral hybrids. Most animals have long stretches of nuclear DNA that are never expressed as proteins and that appear to serve no useful purpose, and although hybridization might be a source of some of this material, it also occurs in animals such as *Drosophila* that seem unlikely to have hybrid ancestries. Such unused DNA of unknown origin would make it difficult to predict how much less should occur in cells of Kirk's species than in cells of brittle-stars with larvae, but there should still be appreciably less. It would be theoretically possible to collect fertilized eggs of Kirk's ophiuromorph from New Zealand beaches and investigate the amount of DNA in their nuclei without investigating the identity of the species, but it is highly desirable that its identity should be established. Incidentally, it is quite irrelevant to my hypothesis whether Kirk's species, the Ophiomyxidae, and the Gorgonocephalidae are primitive or advanced in terms of adult anatomy, although they may be more resistant to cross-fertilization than other ophiuromorphs.

Evidence of the degree of relationship of species in different groups may also be provided by a study of the sequence of genes on molecules of nuclear DNA and cytoplasmic RNA. In the present context, it is obviously important to distinguish among genes that are transcribed into proteins throughout development, those that are transcribed during only part of the developmental process, and those that are never transcribed. It can be predicted, for example, that the genes that are transcribed during the larval development of *Echinus* and *Balanoglossus* will show considerable affinity and those that are transcribed in the adults will show very little. It may also be predicted that molecular analyses of the transcribed genes of larval and adult echinoderms will give very different indications of the relationships between the classes. Such comparisons need not wait until the entire nucleotide sequences of suitable species have been mapped. Raff et al. (1988) have already deduced a phylogenetic tree for living echinoderm classes from the analysis of nucleotide sequences in one form of ribosomal RNA in adult echinoderms. It would be of the greatest interest to make a similar study on larval echinoderms and compare the results with those from adult material. Agreement between the results of the two studies would be a considerable blow to my theories; disagreement would be a considerable boost. It may also be predicted that, in all echinoderms with larvae, the long sequences of genes that are never transcribed will include sequences that show considerable resemblance to those that are expressed in adult enteropneusts, and in most such echinoderms there will be other unused sequences resembling those of other echinoderms, derived from cross-fertilizations within the phylum. Such sequences, both enteropneust and echinoderm, would not be expected to occur in Kirk's ophiuromorph and the Ophiomyxidae and Gorgonocephalidae. It is hoped that it will eventually also be possible to make similar comparisons between the nucleotide sequences of dromioid, anomuran, and brachyuran crabs, between nemertine worms, planarian flatworms, and bryozoans, and between annelids, echiurans, sipunculans, and molluscs, in each case distinguishing those genes that find expression in the embryo and larva (or first larva) from those that are transcribed in adult cells, and from those that appear to be unused.

Botanists have long ago accepted that plant hybrids exist, but whereas crosses between species of the same genus may make up as much as 1% of the world's vascular flora (Stace, 1980), intergeneric hybrids are much less common, and crosses between more distantly related species do not seem to have been considered. The possibility that the effects of hybridity might be shown by early but not late phases in development has also received no attention. Several of the tests used to detect plant hybrids, such as reduced fertility, the variability of the F_2 generation, and distributional evidence, are applicable only to very recent crosses and have little relevance to the cases of supposed hybridization between animals that are put forward in this book. Very often the detection of plant hybrids depends on their characters being intermediate between those of their putative parents. It was noted earlier that crosses between closely related species of animals will also produce intermediate adult and larval characters in the hybrid. Crosses between species in different superfamilies or orders with the same general type of larva seem, however, to have led to intermediate characters of the larvae, although the hybrid adults have retained the characters of their maternal parent. When the putative parents of suspected plant hybrids are known, it is often possible to produce artificial hybrids between these species and compare them with the suspects. There are no cases in the animals under consideration in which the species of both suspected parents of the hybrid ancestor are known, but it may be suggested that the sea-urchin *Arbacia punctulata* is descended from a hybrid between a male *Echinocardium cordatum* and a female of an unknown species of *Arbacia*, and similarly *Sphaerechinus granularis* may be descended from a hybrid between *Echinocyamus pusillus* and a species of *Sphaerechinus*, and *Arbacia lixula* from a hybrid between *Strongylocentrotus franciscanus* and a species of *Arbacia*. In these three cases (considered in Chapter 8) the larva appears to show a mixture of characters of two species but the adult does not. It may, in time, be possible to carry out experimental cross-fertilizations between species closely related to those mentioned and compare the resulting larvae and adults with naturally occurring specimens.

Obviously experimental cross-fertilizations need not be limited to cases that mimic or closely parallel natural cases that are attributable to a hybrid ancestor, for any cross-fertilization in which either or both of the parent species has a larval phase will give some information relevant to the theories under consideration. It may be rather daunting for the experimenter to consider the billions of times eggs and sperm of different species must have come into contact in the sea over hundreds of millions of years, yet all the cases of anomalous embryos and larvae outlined in this book can be explained as the result of no more than 25 successful cross-fertilizations. There may well be other surviving species or groups with hybrids in their ancestries, but, even if the undiscovered cases greatly outnumber those listed here, the total must still be infinitesimally small in relation to the number of potential cross-fertilizations. It must be remembered, however, that (assuming the theory) we see the results of only those cases in which every step in development and metamorphosis has been successful, a viable hybrid has been produced and has reproduced, and its offspring have survived all the rigours of natural selection. The chances of producing a viable hybrid species in the laboratory may be extremely low, but hybrid larvae can be obtained, as is shown later in this chapter, and their progress toward metamorphosis can be studied. The chances of obtaining heterosperm larvae must be vastly greater in the laboratory than in nature, for we do not have to wait for the chance encounter of eggs and sperm that might give a hybrid relevant to the theory. We can select potentially suitable pairs of species, and we can greatly enhance the chances of obtaining the initial cross-fertilization.

The most useful tests of the theory are likely to come from crosses between species that are classified in different superfamilies or are more distantly related. It should be remembered, however, that, in most cases, we are quite ignorant of the amount of genetic difference between taxonomic units, so that, for example, the differences between families in the echinoderms may not be the same as between families in the molluscs, and they may not even be the same between different pairs of families within any one phylum, and the same applies to genera, superfamilies, orders, or any of the other taxonomic units in use. The morphological differences reflected in conventional classifications are, nevertheless, usually the best indications of genetic difference available to us. It was noted in the previous chapter that, in the development of nonhybrids, some phases normally take precedence over others, and it was suggested that a similar system of priorities would probably operate in hybrids to regulate which phase would develop first. This, in some cases, could result in a hybrid that hatched as a paternal larva. When this larval phase was completed, either a paternal or a maternal form might take precedence as the next phase in development, but when there was no clear order of precedence cytoplasmic factors might favour metamorphosis from the paternal-type larva to a maternal-type second larva or juvenile.

Relevant experimental cross-fertilizations may be divided into three categories:

1. It can be predicted that if the two species that are crossed have larvae of the same general type (say, both zoeas or both plutei), then neither larval form will take precedence over the other, and the resulting larva will show a mixture of the larval characters of the two parental species. If, however, this larva manages to metamorphose, it could produce a juvenile resembling the maternal parent. This would give an experimental example comparable to the postulated hybrid ancestor of *Dorhynchus thomsoni* (Chapter 4), and also to the suggested hybrid ancestors of each of the species *Sphaerechinus granularis, Arbacia lixula, Arbacia punctulata, Lytechinus verruculatus,* and *Ophiura texturata* (Chapter 8).

2. If the two species that are crossed have different types of larvae, one type would be expected to take precedence over the other. In general, simpler forms of larvae would be expected to take precedence over more complex forms, and larvae that propel themselves by cilia to take precedence over those that do not, because these criteria seem to apply in all known natural cases in which an animal passes through more than one larval stage. If the paternal larva were to take precedence, we might get a paternal larva, followed by a maternal larva, followed by a maternal juvenile and adult. This, it is suggested, is what happened to produce hybrid species in the ancestry of most modern polychaetes and sipunculans and a number of modern molluscs (Chapter 10). If the maternal larval form took precedence over the paternal, the hybrids would be likely to remain phenotypically maternal throughout life. They would not have new life histories but they would have enhanced genomes. The paternal genes would find no expression in the F_1 hybrids, but we may speculate on whether they

could find expression in F_2 backcrosses with the paternal species. The reverse situation would occur if both the larval and adult paternal body forms took precedence over the maternal larval form: the hybrid would develop as a paternal larva followed by a paternal juvenile and adult. Again we may speculate on the result of backcrossing such forms with the parental species.

3. If the maternal parent has no larva and the paternal parent has a larva, then it may be predicted that usually a paternal-type larva will hatch from the heterozygote, and that, if it metamorphoses, it could go on to produce a maternal-type juvenile. This would parallel the production of the postulated hybrid ancestors of (respectively) all modern dromioid crabs (Chapter 4), nearly all modern echinoderms (Chapters 5–9), all modern echiurans (Chapter 10), and some modern nemertines and bryozoans (Chapter 11). Again one can envisage cases in which the maternal adult takes priority over the paternal larva and adult and the hybrid is phenotypically maternal throughout life, or the paternal larva and adult both take priority over the maternal adult and the hybrid is phenotypically paternal throughout life.

Examples of an experimental cross-fertilization in category 1 and another in category 2 that have already been carried out are outlined below. Both have been sufficiently successful to encourage further experiments on the same lines.

It is beyond the scope of this book to review all the cases of experimental cross-fertilizations and attempted cross-fertilizations that have been carried out in the past. Giudice (1973), however, listed many involving sea-urchins, a group that is particularly relevant to the cases mentioned below. It would seem that, not uncommonly, the mixing of ripe eggs and a very high concentration of active sperm is all that is required to bring about heterosperm fertilization. It is my own experience with four species of sea-urchins that specimens in a spawning condition will usually shed their gametes if they are removed from the water and inverted (mouth upward) over a beaker for up to an hour. After about 30 minutes, spawning may sometimes be stimulated by adding a few drops of seawater, fresh water, or 0.5 molar potassium chloride solution around the mouth or by injecting potassium chloride solution into the coelom. Whichever method is used, the same specimens can be used repeatedly over several weeks.

Many of the experimental crosses involving sea-urchins have been between members of the same genus or the same family, and some have produced larvae with maternal characters, some paternal, and some with a mixture of both. A few have been between more distantly related echinoderms, and these are more relevant to present considerations. Crosses between the regular sea-urchin *Strongylocentrotus purpuratus* and the sand dollar *Dendraster excentricus,* both from the Pacific coast of North America, have been reported by three different authors with different interests: Flickinger (1957), Moore (1957), and Brookbank (1970). Flickinger investigated the production of alkaline phosphatase by the larvae and was, apparently, not very interested in their anatomy. Both Moore and Brookbank compared different features of the hybrids with those of nonhybrid larvae but did not try to raise specimens to the stage of metamorphosis. The two echinomorphs belong to different superorders of the same class, and, as they both have pluteus larvae, the cross comes in category 1 (above). Crosses involving the fertilization of *Strongylocentrotus* eggs with *Dendraster* sperm (S×D) lived longer than the reciprocal crosses (D×S). Hybrids that survived beyond the blastula stage developed more slowly than the larvae of either parent. Flickinger regarded his S×D hybrids as strictly maternal in form but did not describe them. Moore and Brookbank, on the other hand, were agreed that such hybrids varied considerably in form and that some had the general shape of maternal larvae, some of paternal larvae, and some were spheroidal. Brookbank found that all his larvae, regardless of their general shape, had tetraradiate skeletal spicules, like those of paternal larvae rather than the triradiate spicules of maternal larvae, and in many of Moore's hybrids the spicules developed into fenestrated rods, another paternal feature. This apparent inheritance of paternal features of the skeletal rods, even in larvae that were in other respects maternal, is particularly relevant to the cases of *Sphaerechinus granularis, Arbacia lixula,* and *A. punctulata* (above and Chapter 8), the larvae of which all appear to have the skeletal rods of distantly related species, although the general shape of the larvae is consistent with the generic classifications of their corresponding adults. The slow development of the experimental hybrids brings to mind the larvae of *Lytechinus verruculatus,* which show marked differences in both form and skeleton from other species of *Lytechinus* and which also develop unusually slowly (Chapter 8). The anomalies of *L. verruculatus* are fully consistent with the suggestion that it has a hybrid in its recent ancestry, that most of the larval characters are inherited from the male parent of this hybrid, and all the adult characters from the female parent.

Brookbank observed the formation of enterocoelic pouches in his *Strongylocentrotus* × *Dendraster* hybrid larvae. These larvae were kept alive for up to 8 days, but this was not long enough for juvenile rudiments to appear. Some of Flickinger's larvae "were kept as long as 12 days after fertilization and they were still perfectly healthy at that time," but he made no mention of their morphological development. Mortensen (1921) found that, in the laboratory, the juvenile would start to form in different (nonhybrid) species of planktotrophic echinopluteus larvae in anything from 13 to 35 days after fertilization, and some that fed only on internal yolk could complete metamorphosis in as little as 4 or 5 days. We know that *Strongylocentrotus* × *Dendraster* larvae develop slowly, and we cannot predict whether any would be capable of metamorphosis, but there seems to be a good chance that some could be kept alive at least until the beginning of the formation of the juvenile. It would be of the utmost interest to the hypotheses now under consideration to know what form that juvenile would take.

A laboratory cross-fertilization in category 2 (above), which I have carried out several times, gives a very clear indication that a larval form can be transferred from one animal to a very distantly related one by cross-fertilization. This is the fertilization of eggs of ascidians with sea-urchin sperm. Further experiments are still in progress, and a full report will be submitted for publication to a scientific journal when they are completed. In the meantime, however, I give the following account of the results to date.

Ascidians, often called sea squirts, are usually classified as members of the subphylum Urochordata of the phylum Chordata. We belong to the subphylum Vertebrata of the same phylum. Adult ascidians (Fig. 13.1) are sedentary, sac-like, hermaphrodites in which an inhalent siphon leads to a highly perforated pharynx, while an exhalent aperture carries away faeces, excretory products, sperm, and, usually, eggs, but some brood the eggs internally. Some have no larvae, but the majority hatch as small tadpoles (Fig. 13.1) that swim for anything from a few hours to several days before settling and metamorphosing. These tadpole larvae are fully formed at the time of hatching and are extremely different from the ciliated blastulae that hatch from sea-urchin eggs or the pluteus larvae into which such blastulae develop.

Eggs of the ascidian *Clavelina lepadiformis* were used in only one experiment. A specimen was found on a submerged rope, and it shed eggs overnight in a beaker. (The eggs of this species are frequently retained inside the outer test until after fertilization, and the tadpoles make their way out through the exhalent aperture. The stress of handling and temperature change probably led to the complete discharge of the eggs in this case.) A further group of about 25 experiments involved the large ascidian, *Ascidia mentula*, conveniently found growing on the walls of a covered seawater storage tank at the Port Erin Marine Laboratory, and eggs were obtained by gently squeezing adult specimens. The sea-urchin

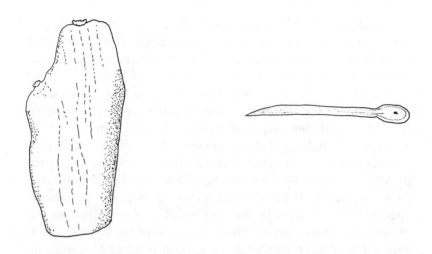

FIG. 13.1. An adult *Ascidia mentula* (about 7 cm high) and its tadpole larva (about 0.5 mm long).

used was *Echinus esculentus,* which also occurs near the Port Erin Laboratory, and sperm was obtained by the methods mentioned earlier in this chapter. The highest concentrations of sperm, however, came from four specimens that had already started to spawn spontaneously in a holding tank.

Of the 23 attempts to cross-fertilize ascidian eggs with *Echinus* sperm in 1988 and 1989, 17 produced no larvae. In three cases tadpole larvae emerged, perhaps as a result of the *Echinus* sperm stimulating cell division without playing any further part in the production of the resulting larva. In two of these cases in which some eggs hatched as tadpole larvae, however, others hatched as ciliated blastulae, and in another four cases all the eggs that hatched did so as ciliated blastulae. In four of these experiments the blastulae developed further. The larvae obtained from eggs of *Clavelina* developed as far as gastrulae, but they remained spheroidal and died after 7–10 days. Larvae obtained from eggs of *Ascidia,* however, developed beyond the gastrula stage in three cases, producing, respectively, 1, over 200, and 1 undoubted pluteus larvae, with slender arms supported by calcareous rods. These were fed on cultured flagellates. In all experiments the sperm suspension used was sufficiently concentrated to look milky to the naked eye, and in that which produced over 200 pluteus larvae it was of a creamy consistency. About 75% of the hybrid pluteus larvae obtained were stunted or showed some deformity, such as one or more of the arms failing to develop (Fig. 13.2), but this may well have been a result of more than one sperm reaching the egg nucleus, a condition known as polyspermy. Similar stunted or deformed larvae can be produced by fertilizing *Echinus* eggs with fairly concentrated *Echinus* sperm, although eggs exposed to milky or creamy sperm, like that used in the successful cross-fertilization experiments, do not hatch. The other 25% of the hybrid larvae showed no deformities (Fig. 13.3, Frontispiece) and were indistinguishable from normal *Echinus* larvae. In these experiments some larvae survived for 8 weeks or more, but none metamorphosed or showed any sign of metamorphosis. None of the nonhybrid *Echinus* larvae, reared as controls, metamorphosed either.

Sometimes larvae fail to metamorphose, and this is true whether they result from homosperm or heterosperm fertilizations. In 1990, however, a further cross-fertilization of *Ascidia* eggs with *Echinus* sperm produced larvae which did metamorphose. In this experiment over 3000 pluteus larvae were obtained, and diatoms were provided as food. More than 70 larvae developed juvenile *Echinus* rudiments (Fig. 13.4d) and 20 successfully metamorphosed in 37–50 days at 18°C (Fig. 13.4e). They were placed in a small aquarium tank at 15°C, and 1 year after fertilization, four are surviving as apparently healthy urchins. They range in diameter from about 0.6 to 2.0 cm, and they are morphologically indistinguishable from nonhybrid *Echinus*. These specimens will provide the opportunity to remove one or two tube-feet from each to investigate whether their ascidian chromosomes have remained intact and recognizable, as well as their echinoderm chromosomes. If they survive to maturity, I hope it will be possible to breed from them, crossing one hybrid urchin with

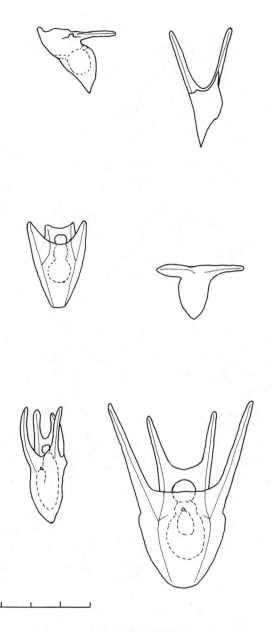

FIG. 13.2. Variation among 8-day-old hybrid larvae produced by fertilizing *Ascidia* eggs with *Echinus* sperm. The scale represents 0.5 mm.

A

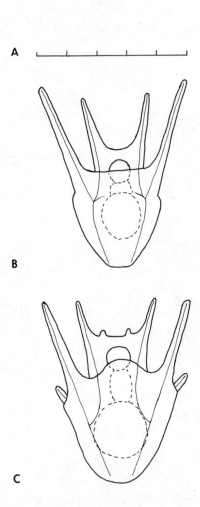

B

C

FIG. 13.3. Examples of hybrid larvae produced by the fertilization of *Ascidia* eggs with *Echinus* sperm (experiment conducted in 1989). (a) Three days from fertilization; (b) 8 days; (c) 15 days; (d) 22 days; (e) 29 days. The scale represents 0.5 mm.

LARVAE AND EVOLUTION

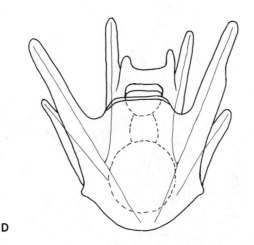

D

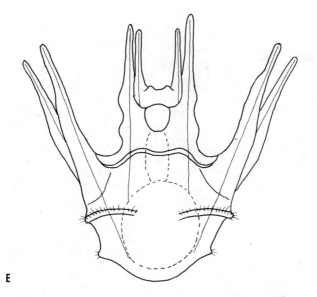

E

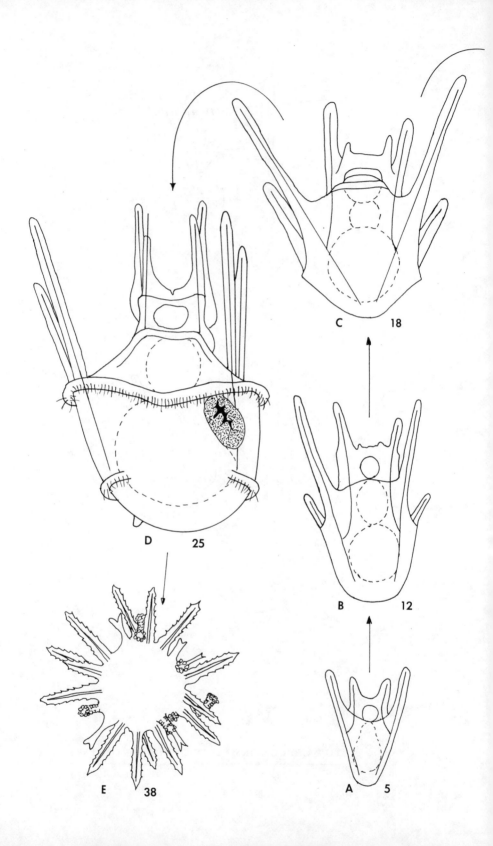

C 18

D 25

B 12

E 38

A 5

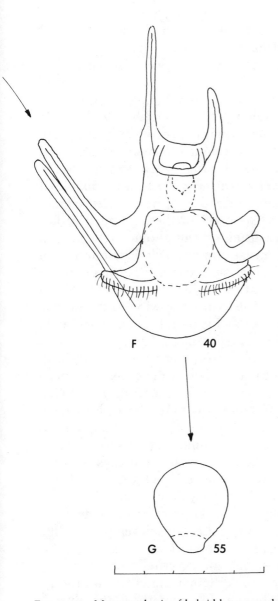

FIG. 13.4. Metamorphosis of hybrid larvae produced by the fertilization of *Ascidia* eggs with *Echinus* sperm (experiment conducted in 1990). All larvae went through a pluteus phase (a–c), but, although some developed *Echinus* rudiments (d) and settled as small sea-urchins (e), others resorbed their pluteal arms (f) and settled as small spheroids (g). Ages are given in days from fertilization. The scale represents 0.5 mm.

another (if they are of different sexes) and crossing hybrid urchins with nonhybrid *Echinus* and with *Ascidia*.

The pluteus larvae that developed *Echinus* rudiments were, however, only a minority. Many larvae, not unexpectedly, died without attaining full size or developing the full complement of arms and epaulettes, but several hundred did attain this state of development without producing any trace of *Echinus* rudiments. They continued to swim actively, and gradually resorbed their pluteal arms, often asymmetrically (Fig. 13.4f). At 18°C, the first of these larvae was capable of settlement at 57 days from fertilization. It had condensed into a spheroid of ciliated cells with a smaller, rounded protuberance at one end, measuring overall about 0.3 × 0.2 mm (Fig. 13.4g). Another 20 larvae attained a similar state within the next 16 days. Each larva could change the shape of the main body from spheroidal to ovoidal, and could fix itself very firmly to the walls of the rearing bowl with thread-like organs of attachment put out by the protuberance. They could release themselves, swim, and reattach. These larvae were placed in a small aquarium at 15°C, but none has survived.

At this stage we do not know whether these small, rounded larvae are capable of developing further, or, if they are, what they will develop into. We do know that, although they developed from pluteus larvae, they are quite unlike any known stage in the development of an echinoderm. It should be remembered that they developed from ascidian eggs, so that, if their maternal genes have remained intact, they could conceivably develop into ascidians. They do not look like tailless tadpoles, but the tadpole may not be an essential stage in the development of ascidians. Some ascidians develop without a tadpole stage, and, where a tadpole does occur in the life history, its metamorphosis has all the criteria of a larva acquired from another group. Although the larva and the juvenile ascidian are both bilaterally symmetrical, the juvenile does not adopt the orientation of the larva, and most of the larval tissues and organs are discarded at metamorphosis (Brien, 1948). Cloney (1978) confirmed earlier reports of the larval and future adult nervous systems existing side by side or one above the other in developing ascidians, before the larval system disintegrates. I shall return to the question of the origin of the ascidian tadpole larva in the final chapter. The possible further development of the rounded larva and the genetic composition of all hybrid larvae pose intriguing questions that can be answered only by further research. Fortunately, as I have shown, some hybrid larvae at least are not difficult to obtain.

There is no evidence that sea squirts and sea-urchins have ever hybridized in nature, and crossing them experimentally does not prove that an ancestor of any animal group ever acquired a larval form by cross-fertilization. The experiments show, however, that the first step in the process of transferring a larval form from one species to a very distantly related one by cross-fertilization is entirely feasible. It shows that a heterozygote can give rise to a paternal-type larva rather than to a chimera. The experiments also show that a hybrid larva is capable of metamorphosis, although, in this case, all the juveniles were of the paternal form. Perhaps the echinoderm juvenile and adult body always take precedence over both the tadpole larva and the adult of a urochordate. I suggest that the body form of a juvenile echinoderm took precedence over that of a juvenile hemichordate in the hybridization that produced the first echinoderm with bilateral larvae.

Crosses in category 3, in which the eggs of a species without a larval phase are fertilized with sperm from a species with one, may also be attempted. The female parents may be drawn either from animals that have apparently never had a larva, such as the Chaetognatha, or from others that have secondarily adopted direct development, examples of which occur in many phyla. Fertilization in the Chaetognatha is internal, so, if animals from this group are used, the experimenter has the problem of introducing the foreign sperm into the body of the arrow worm or of extracting undamaged eggs. The possibility of using crustaceans in such experiments should not be overlooked. Male shrimps and lobsters may be induced to eject sperm by giving them an electric shock of 3–9 volts AC, depending on size. Aiken et al. (1984) used 9 volts for the lobster *Homarus americanus,* but I have found 3 volts sufficient for the shrimp *Palaemon elegans.* The amphipod crustacean *Gammarus duebeni* is an example of a species with no larval phase whose females will lay eggs even when no male is present.

It is hoped that many biologists will try to produce hybrids to test the theory that has been put forward. It would obviously increase the scope of such experiments if sperm from several species were to be stored at low temperatures. This would ensure that sperm from these species was always available, and it would permit crosses between animals that normally breed at different seasons. Either sperm or eggs at 0–5°C, in insulated containers, may be flown to different parts of the world for use in such experiments.

IV
CONCLUSIONS

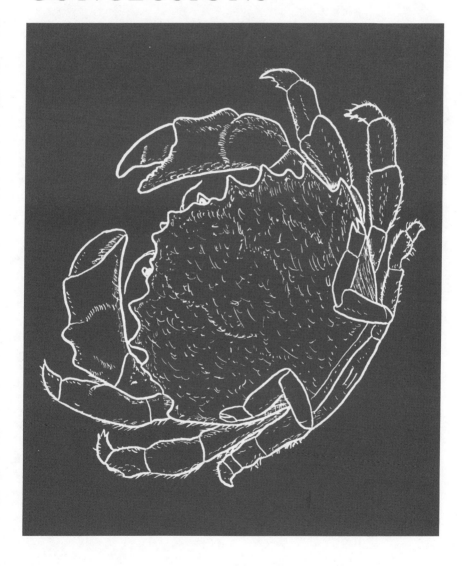

14
Toward a New Zoology

Problems of classification if present theories accepted—New life-histories acquired by horizontal transfer may have led to accelerated evolution— Possibility of horizontal transfer affecting adults, leading to 'hopeful monsters' —Urochordates present possibility of adult of one group having become larva of another after horizontal transfer—Suggested reclassifications of some animal groups—Need for broader debate on principles of phylogeny and taxonomy

This book is about one aspect of animal development and evolution, but there are several other aspects of this broad theme that it either ignores or barely discusses. It seeks to explain a number of anomalies in larval form and methods of metamorphosis, but it adds nothing to existing theories on the origins of larvae. It emphasizes the evolutionary distinction between radial and bilateral symmetry but adds nothing to our knowledge of the origins of either. It opens up a new aspect of ontogeny and phylogeny that has very little to do with neoteny, paedomorphosis, and recapitulation. I have tried to show that there are a great many examples of larvae and a smaller number of examples of embryos that, in terms of neo-Darwinian evolutionary theory, seem out of place with respect to their corresponding adults. These anomalies are, however, all consistent with the concepts that embryonic and larval forms, each of which evolved in one lineage of animals, have sometimes appeared at later dates in other lines, that these transfers of types of development have been the result of transfers of genetic material, and that these gene transfers have probably resulted from heterosperm fertilizations. Some of the consequences of adopting these concepts will now be examined. In doing so, however, it must be emphasized that, although an idea must be rejected if its implications are not reconcilable with reality, it should not be rejected merely because these implications are inconvenient.

The new concepts introduced in this book are regarded as additions to Darwin's view of organic evolution as the gradual accumulation of variations, not as substitutes for it. I have assumed that both adult and larval forms have evolved in a neo-Darwinian manner and that larvae that have been transferred to new evolutionary lines have continued to evolve as parts of new life histories. My new hypotheses, however, conflict with Darwin's views in two important respects. The sudden introduction of a new larval form into a life history would have been regarded by Darwin as a "saltation," a major leap that played no part in his perception of evolution. This type of saltation involves the amalgamation of two genomes that have themselves evolved in a Darwinian manner. It should be distinguished from a large mutational change or macromutation, which is another form of evolutionary saltation, either actual or hypothetical.

I also disagree with Darwin's assumption that a species can have only one lineage, for I contend that a larval form can originate in one group of animals and later appear in another. Most biologists since Darwin have retained his assumption as far as animals are concerned, but the notion of viable hybrids is now widely accepted in the fields of both cell biology and botany. As a result of the efforts of Lynn Margulis, there is increasing recognition that symbiotic associations between distantly related bacteria produced cells that have evolved as proctists, fungi, plants, and animals. Such cells are, therefore, hybrids from the genetic point of view, although the hybridization, in this case, did not involve cross-fertilization. It has long been recognized that a number of plants will hybridize in the laboratory and in nature, and the International Code of Botanical Nomenclature (ICBN, 1972 and earlier editions) includes rules and recommendations for naming hybrids. The Code, however, considers only hybrids between relatively closely related plants, and botanists have not considered the possibility of different developmental phases of the same species showing the characters of different taxa. Perhaps they should. The algae are a group covered by the International Code of Botanical Nomenclature, although, in the classification of Margulis (1981), they are included in the Protoctista rather than the Plantae. My colleague, Trevor Norton, has drawn my attention to the zoospores of the green alga *Derbesia marina,* which look not unlike trochophores, although there is no mouth or alimentary canal. No other alga with similar spores is known. It may eventually be shown that *Derbesia* has animal as well as protoctist genes on its chromosomes, indicating that genetic transfer may take place between kingdoms.

I am trying to get biologists to consider the notion that some species may have early developmental phases that show either the characters of one group (A) or a mixture of two groups (A and B), whereas their later phases show the characters of group B only, and that their genomes may contain genes from both group A and group B. As a result of genetic engineering, we are becoming accustomed to the idea of a chimera that shows a mixture of characters of two distinct species simultaneously. I now claim that there is evidence that most echinoderms are sequential chimeras, which change phylum during development, and comparable sequential chimeras occur in several other phyla. The concept of an individual that changes from one phylum to another during its life history poses profound taxonomic problems at all levels from individual animals to phyla. These problems are discussed in a recent essay (Williamson, 1991), but any solution must involve radical rethinking of how we regard individuals, species and phyla.

In some groups the life histories of all the species may show evidence of descent from a hybrid ancestor, in others most but not all species are affected, and in yet others only a minority. This means that not only may we require separate classifications for the early and late phases in development and separate dendrograms to show their ancestries but such separations may be necessary for only some of the members of the adult group. Ideally we require a system of classification of living and extinct animals that, to take one example, can reflect the probability that all modern dromioid crabs have a genome with both anomuran and brachyuran elements that expresses itself as anomuran larvae and brachyuran adults, whereas the earliest dromioids were wholly brachyuran, both genetically and phenotypically. This ideal classification should also take into consideration that the great majority of modern echinoderms seem to have genetic affinities with enteropneusts, but these are expressed only in the embryonic and larval stages, not in the adults. Such genetic affinities were acquired at different times by different groups within the phylum, and a few echinoderms never acquired them. It is even possible that some members of one or more extant species will have acquired larvae whereas other members have not, a condition that, I hold, must have occurred many times in the past. I sympathize with taxonomists who might shrink from contemplating such complex problems, but I did not create the problems. I have merely exposed them. I repeat that a hypothesis should not be rejected merely because its implications are inconvenient.

Acceptance that larvae can be transferred between distantly related groups does not invalidate the use of larval characters in all classifications. When animals can be arranged in groups that are compatible with both adult and larval characters, this provides a strong indication that the adults and larvae concerned have evolved together, as different phases in the same life history, and the resulting classification is more likely to provide a true reflection of phylogeny than one that ignores the larvae. Thus Rice (1980) was able to review the classification of brachyuran crabs in light of larval characters, and the present proposal that a very few such crabs owe some or all of their larval characters to genetic transfers from distantly related groups does not detract from the value of Rice's work.

If a group of animals can acquire a larval form where there was none previously or gain an additional larval form at the beginning of the life history, this might have considerable bearing on the evolution of adult form within the group concerned, even though the adults may not be directly affected by the genetic transfer responsible for the introduced larval phase. The newly acquired larvae could greatly increase the efficiency of dispersal, allowing animals with such larvae to spread to areas of reduced competition and to new ecological niches. This could result in bursts of evolutionary diversification of adult form without necessarily requiring a mechanism of evolution other than the gradual accumulation of mutations. Such bursts, or "punctuated equilibria" as Eldredge and Gould (1972) have called them, are common features of the fossil record, and I suggest that cross-fertilizations, resulting in the transfer of larval forms, might have been responsible for at least some of them.

In all the examples of apparent hybrids and their descendants that I have considered, I have found only very limited evidence of changes in the form of adult animals that might have resulted from the addition of new genetic material. I have suggested that the partial bilateral symmetry of adult holothurians may be the result of genetic factors that produce bilateral symmetry in the larvae continuing to exert some influence in the adults. The genetic recipe for these bilateral larvae was, I believe, originally transferred from an enteropneust, so, to that extent, the bilateral features of adult holothurians may be regarded as a feature transferred from an enteropneust. If, however, I am right in thinking that the genetic transfer was the result of cross-fertilization, the original hybrid would have had the genetic recipe for not only an enteropneust larva and an echinoderm adult but also for an enteropneust adult. I have given reasons in Chapter 12 why I think this hybrid matured as an echinoderm rather than as an enteropneust, and, in general, the genetic prescription for the adult enteropneust seems to have been lost, or has remained totally unused in all the descendants of the hybrid. The partial bilateral symmetry of adult holothurians, however, may be interpreted as a character of adult, rather than larval, enteropneusts. This suggests that, in some circumstances, some adult features of the male parent of the original hybrid may eventually reappear, although in this case the suggestion is extremely tentative.

Another group also provides evidence that may be interpreted in the same way, although in this case the evidence is stronger. This group is the Echiura (discussed in Chapter 10), in which transient metameric segmentation can appear immediately after metamorphosis, affecting different tissues in different genera. It was suggested that the Echiura acquired their trochophore larval phase from a polychaetous annelid and that segmentation is a character of adult annelids that may make a brief appearance in juvenile echiurans but is quickly suppressed. This example considerably strengthens the suggestion that genes that came from the male in the original hybrid can, in some circumstances, be expressed after metamorphosis. Metameric segmentation is brief in the known Echiura, but must the influence of such paternal genes always be brief? Indeed, could not the chaetae of echiurans be an adult character that the group acquired from the polychaetes at the same time as it acquired trochophore larvae? The adult morphologies of some groups are probably so different that any attempt to mix them is doomed to failure, but can we deny the possibility that any group had its origin in a hybrid between two different groups? The present evidence that hybridization has taken place during the course of evolution comes from the distribution of types of early development in the animal kingdom, but, if this evidence is accepted, could not the same process have produced the "hopeful monsters" envisaged by Goldschmidt (1940) and the sudden initiation of new evolutionary lines that appear to punctuate the general equilibrium of the fossil record (Eldredge and Gould, 1972)?

I have been pursuing the argument that many changes in the form of early development have been brought about by cross-fertilizations between distantly related animals, but there seems no reason why cross-fertilizations should be restricted to cases in which new embryos or new larvae are introduced.

If the male parent of a hybrid had no larva, the life history of the hybrid would be expected to be the same as that of its female parent, whether or not she had a larva, but again the unused male genes might not always remain unused indefinitely. A hybrid with a maternal life history would also seem to be the most likely outcome if both parents had larvae but that of the female took developmental precedence. This might happen, for example, if eggs of the sea-urchin *Echinus* were fertilized with sperm of the ascidian *Ascidia,* the reciprocal cross to that described in the previous chapter. The possibility that the hybrid would progress from a maternal larva to a paternal larva and then to a maternal juvenile and adult seems less likely, and the chances of its successfully carrying out the two metamorphoses involved would be extremely remote.

The fertilization of *Ascidia* eggs with *Echinus* sperm, referred to in the previous chapter, seems to provide a case of the genes from the male parent taking precedence not only during embryonic and larval development but also at the time of metamorphosis, resulting in a hybrid with a wholly paternal life history. (This is leaving aside the rounded larvae into which some hybrid plutei develop, larvae that do not occur in the paternal or maternal life history.) In cases such as this, however, there could be a conflict after metamorphosis between the genetic precedence favouring expression of the paternal genes and the cytoplasmic factors favouring expression of the maternal genes, and perhaps, in some cases, both would find some expression in the adult animal. Whether or not this happens in the later development of the *Ascidia* × *Echinus* hybrids, some cross-fertilizations in which the male body form takes precedence both before and (initially) after metamorphosis could perhaps lead to hopeful monsters by some characters of the female parent appearing in later development. It is outside the scope of this book to try to review the adult form of all extant and extinct organisms to consider which, if any, might have arisen as hybrids, but it must be pointed out that hybridization seems to provide a possible mechanism for the production of morphological saltations. It is hoped that the search for hopeful monsters will continue more hopefully.

Evidence, or so I interpret it, for the transfer of larval and embryonic form from one animal lineage to another is presented in the first 11 chapters of this book. Here, in this chapter, I have briefly considered the possibility that the implied genetic transfers may have also occasionally affected adult form. Now I wish to draw attention to a case in which the adult form of an animal seems to have been transferred to another lineage, appearing in this second line as a larva. This intriguing case concerns the animals usually grouped together as the subphylum Urochordata of the phylum Chordata, the phylum to which we belong. The Urochordata is made up of the three classes Ascidiacea or Tunicata (ascidians), Thaliacea (salps, pyrosomas and doliolids), and Appendicularia or Larvacea (appendicularians). The cross-fertilization of eggs of a solitary ascidian (*Ascidia mentula*) with sea-urchin sperm was described in the previous chapter. The Ascidiacea also includes colonial species, in which groups of individuals share a common atrial opening. Thaliaceans are also either solitary or colonial. They have body forms somewhat resembling those of ascidians, but although all ascidians are sedentary, all thaliaceans are planktonic. Appendicularians are also planktonic but, as adults, have a totally different body form. Most ascidians and doliolids have tadpole larvae, although other thaliaceans do not. Appendicularians, on the other hand, mature as tadpoles.

The process of metamorphosis from a tadpole larva to an adult ascidian is truly cataclysmic, as I have already mentioned in Chapter 13, and doliolids probably undergo a similar metamorphosis. This cataclysmic metamorphosis is evidence, in my view, that the larva originated in another group, and, if this is so, the other group is rather obvious. Appendicularians are classified with ascidians and thaliaceans because of the similar tadpole form that occurs in all or some members of each group, whether as adult or larva. I suggest that sperm of an ancestor of the modern appendicularians probably fertilized an egg of an ancestor of the modern ascidians, producing a tadpole larva that metamorphosed into a juvenile ascidian. Probably a separate cross-fertilization led to a tadpole larva that metamorphosed into a juvenile doliolid, but, if the lineages of doliolids and ascidians separated after those of salps, pyrosomas, and ascidians, one successful cross-fertilization would have been sufficient. This means that, according to my hypothesis, the adult form of one group became the larval form of another, or of two others.

The suggestion that, throughout their evolutionary history, the appendicularians never metamorphosed into any other body form was first proposed by Haeckel (1874) and later adopted by Lohmann (1933) (although these authors did not suggest that this body form became an ascidian larva as the result of a cross-fertilization). An alternative theory was proposed by Willey (1893) and adopted and elaborated by Garstang (1928), and this is more generally accepted today. Under this view, the appendicularians originally metamorphosed from tadpoles into a tailless adults, and they probably shared their ancestry with the modern doliolids. The modern appendicularians then arose by a process of paedogenesis, maturing in what was previously the larval state. I hope that my indication of a possible method of uniting the tadpole body form with that of an ascidian or a doliolid will lead to a revival of the Haeckel–Lohman theory that the tadpole-like adult was primitive. My view goes further in suggesting separate origins for tadpoles on the one hand and ascidians and thaliaceans on the other. Under either of these views, the tadpole-like appendicularians must be regarded as closer to the ancestry of the vertebrates than they are under the Willey–Garstang theory. It must be borne in mind, however, that the adult body forms of ascidians and thaliaceans show some chordate affinities (Jefferies, 1986). Adoption of my views may therefore lead to some reassessment of ideas on the origin of vertebrates, but the immediate effect will be to separate the adult forms of those groups currently regarded as urochordates. The name, urochordate, should be restricted to the tadpole form, with a notochord in its tail, and the adult form of ascidians and thaliaceans should be united under a different name.

In the past, similar embryos and similar larvae have always been regarded as indications of phylogenetic affinity between the groups in which they occur, but if, as is now claimed, embryonic and larval forms can be transferred between animals in different lineages then embryos and larvae may give no reliable indication of the paths along which adult animals have evolved. Thus the long-established grouping of echinoderms with hemichordates and chordates is now challenged, but the affinity between echinoderm larvae and hemichordate larvae is upheld. This means that, as suggested some years ago by Dr. Isaac Frost (personal communication), echinoderm embryos and larvae could be invaluable as test animals for drugs being considered for medical and veterinary use, for chemicals that produce teratogenic effects on the development of one enterocoelous deuterostome could well produce comparable effects on the development of others. Echinoderm larvae are frequently easy to rear, and their well-defined shapes should facilitate the detection of drug-induced abnormalities.

It is, of course, not only the relationship between adult echinoderms and chordates that is called in question. It is also suggested that the annelids, echiurans, sipunculans, and molluscs have evolved quite independently of each other, in spite of similar larvae occurring in at least some representatives of each of these phyla, and the same applies to the flatworms, nemertines, and bryozoans. There is no reason to question the assumption of the nineteenth-century biologists that protostomy and schizocoely on the one hand and deuterostomy and enterocoely on the other are fundamentally different methods of development, but if one method can supplant the other then they are no more reliable than larvae as indicators of adult affinity. For the past century, embryonic and larval clues to the relationships between animal groups have been accepted as the basis for grouping phyla into superphyla and for constructing phylogenetic trees such as that shown in Figure 3.1. The theories now being put forward, however, imply that these clues have been misinterpreted and that the widely accepted groupings and trees are totally misleading. Until more direct evidence of genetic affinity is available, new groupings of adult animals will have to be based almost entirely on adult characters, and in some cases this will resolve old conflicts. Bryozoans and hemichordates, for example, can now be grouped together as lophophorate animals, and the fact that bryozoans develop as protostomes whereas hemichordates develop as deuterostomes need not be regarded as conflicting with this classification. There is, however, the further complication that, if hybridizations have affected adult as well as larval form, perhaps the lophophore has been transferred between distantly related groups.

The rejection of larval clues as an infallible guide to ancestry gives no justification for reviving the Articulata, Annulosa, or Appendiculata as a group for most segmented animals. Annelids are animals that have evolved a septate coelom as an aid to efficient burrowing (Clark, 1964), whereas arthropods have evolved an articulated exoskeleton to combine protection and mobility. I know of no arthropod with a segmented coelom, and I question whether the two forms of metamerism had a common origin.

Ultimately it is hoped that it will be possible to classify animals, and all organisms, according to their genetic affinities. It will, however, be important to distinguish between genes that are transcribed in embryos and larvae, those that are transcribed in juveniles and adults, and those that are never transcribed. In most of the cases considered in this book, the entire genetic prescriptions for larval form seem likely to have originated in another evolutionary line, although we cannot rule out the possibility of introduced genetic material affecting adult form and occurring in untranscribed portions of the genome.

There has been a lively debate on the methods of classification for some years, but this has been largely concerned with the merits and demerits of cladistic analysis. I am now asking that this debate should be widened to cover the basic principles of phylogeny. Lynn Margulis has already shown that the amalgamation of widely different organisms has played an essential role in the evolutionary origin and history of cells and the organelles within them and protruding from them. I am now asking biologists to consider the evidence that hybridization between animals in different phylogenetic lines has also played and is playing a major part in shaping the form of embryonic and larval development of many groups. The evidence is, I submit, considerable, and several tests of the theory have been suggested. The suggestion that the same process might have produced saltations in adult morphology is not backed by the same weight of evidence, but it should certainly be explored further. The acceptance that symbiotic associations between bacteria have produced cells and that hybridizations between distantly related animals have produced new life histories obviously involves the acceptance that evolution cannot be explained wholly in terms of gradual descent with modification and natural selection. It does not, however, denigrate or disparage Darwin's view of evolution any more than relatively denigrates or disparages Newtonian physics. Without the original inspired theories it would have been impossible to add the later modifications.

Bibliography

Aiken, D.E., Waddy, S.L., Moreland, K., and Polar, S.M. 1984. Electrically induced ejaculation and artificial insemination of the American lobster *Homarus americanus*. *Journal of Crustacean Biology 4:* 519–527.

Anderson, D.T. 1973. *Embryology and Phylogeny in Annelids and Arthropods*. Pergamon Press, Oxford.

Baker, A.N., Rowe, F.W.E., and Clark, H.E.S. 1986. A new class of Echinodermata from New Zealand. *Nature, London 321:* 862–863.

Barnes, R.D. 1980. *Invertebrate Zoology*, 4th ed. Saunders, Philadelphia.

Barnes, R.S.K., Callow, P., and Olive, P.J.W. 1988. *The Invertebrates: A New Synthesis*. Blackwell Scientific Publications, Oxford.

Bell, F.J. 1892. *Catalogue of the British Echinoderms in the British Museum (Natural History)*. British Museum, London.

Borradaile, L.A., Potts, F.A., Eastham, L.E.S., and Saunders, J.T. 1935. *The Invertebrata*. Cambridge University Press, Cambridge.

Brien, P. 1948. Embrachement des Tuniciers. Pp. 553–894 in Grassé, P.P. (ed.), *Traité de Zoologie, Anatomie, Systématique, Biologie, Tome XI, Échinodermes, Stomocordés, Procordés*. Masson, Paris.

Brien, P. 1960. Classe de Bryozoaires. Pp. 1053–1335 in Grassé, P.P. (ed.), *Traité de Zoologie, Anatomie, Systématique, Biologie, Tome V, Fasc. II, Bryozoaires, Brachyopodes, Chétognathes, Pogonophores, Mollusques Aplacophres, Polyplacophores, Monoplacophores, Lamellibranches*. Masson, Paris.

Brookbank, J.W. 1970. DNA synthesis and development in reciprocal interordinal hybrids of a sea-urchin and a sand dollar. *Developmental Biology 21:* 29–47.

Chadwick, H.C. 1914. Echinoderm larvae. *L.M.B.C. Memoirs 22:* 1–32.

Cherfas, J. 1982. *Man Made Life. A Genetic Engineering Primer*. Blackwell, Oxford.

Chia, F.S., and Burke, R.D. 1978. Echinoderm metamorphosis: Fate of larval structures. Pp. 219–234 in Chia, F.S., and Rice, M.E. (eds.), *Settlement and Metamorphosis in Marine Invertebrate Larvae*. Elsevier, New York.

Clark, R.B. 1964. *Dynamics in Metazoan Evolution. The Origin of the Coelom and Segments.* Oxford University Press, Oxford.

Clarkson, E.N.K. 1979. *Invertebrate Palaentology and Evolution.* Allen and Unwin, London.

Cloney, R.A. 1978. Ascidian metamorphosis: review and analysis. Pp. 255–282 in Chia, F.S., and Rice, M.E. (eds.), *Settlement and Metamorphosis in Marine Invertebrate Larvae.* Elsevier, New York.

Darwin, C. 1859. *The Origin of Species by Means of Natural Selection or the Preservation of Favoured Races in the Struggle for Life.* John Murray, London. (Reprinted 1985, Penguin Classics, London.)

Dawkins, R. 1986. *The Blind Watchmaker.* Longmans Scientific & Technical, Harlow, Essex, England.

Dawydoff, C. 1959. Classe des Echiuriens. Pp. 855–907 in Grassé, P.P. (ed.), *Traité de Zoologie, Anatomie, Systématique, Biologie, Tome V, Fasc. I, Annélides, Myzostomides, Sipunculiens, Echiuriens, Priapuliens, Endoproctes, Phoronidiens.* Masson, Paris.

De Beer, G.R. 1930. *Embryology and Evolution.* Oxford University Press, Cambridge.

De Beer, G.R. 1940, 1951, 1958. *Embryos and Ancestors.* (Revised editions of De Beer, 1930.) Oxford University Press, Oxford.

Eldredge, N., and Gould, S.J. 1972. Punctuated equilibria: an alternative to phyletic gradualism. Pp. 82–115 in Schopf, T.J.M., and Thomas, J.M. (eds.), *Models in Paleobiology.* Freeman Cooper, San Francisco.

Fell, H.B. 1941. The direct development of a New Zealand ophiuroid. *Quarterly Journal of Microscopical Science 82:* 377–441, 3 pls.

Fell, H.B. 1948. Echinoderm embryology and the origin of chordates. *Biological Reviews 23:* 81–107.

Fell, H.B. 1963. The phylogeny of sea-stars. *Philosophical Transactions of the Royal Society of London, Series B 246:* 381–435.

Fell, H.B. 1968. Echinoderm ontogeny. Pp. S 60–S 85 in Moore, R.C. (ed.), *Treatise on Invertebrate Paleontology, Part 5, Echinodermata.* Geological Society of America and University of Kansas Press, Lawrence, Kansas.

Fernando, W. 1931. The origin of the mesoderm in the gastropod *Viviparus* (= *Paludina*). *Proceedings of the Royal Society of London, Series B 107:* 381–390.

Flickinger, R.A. 1957. Evidence from sea-urchin–sand dollar hybrid embryos for a nuclear control of alkaline phosphatase activity. *Biological Bulletin 112:* 21–27.

*Frey, H., and Leuckart, R. 1847. Lehrbuch der Anatomie der wirbellosen Thiere, in Wagner, R. *Lehrbuch der Zootomie.* Thiele 2.

†Garstang, W. 1928. The morphology of the Tunicata and its bearing on

*Cited from Hyman (1940–59); †cited from Jefferies (1986).

the phylogeny of the Chordata. *Quarterly Journal of Microscopical Science 72:* 51–187.

Garstang, W. 1966. *Larval Forms and other Zoological Verses.* Blackwell, Oxford.

Gibbs, P.E. 1977. *British Sipunculans* (Synopses of the British Fauna 12). Academic Press for the Linnean Society of London.

Giudice, G. 1973. *Developmental Biology of the Sea-Urchin Embryo.* Academic Press, New York.

Goad, L.J., Rubinstein, I., and Smith, A.G. 1972. The sterols of echinoderms. *Proceedings of the Royal Society of London, Series B 180:* 223–246.

Goldschmidt, R. 1940. *The Material Basis of Evolution.* Yale University Press, New Haven, Connecticut.

Gontcharoff, M. 1961. Embranchement des Némertiens (Nemertini G. Cuvier 1817; Rhynchocoela M. Schulze 1851). Pp. 783–886 in Grassé, P.P. (ed.), *Traité de Zoologie, Anatomie, Systématique, Biologie. Tome IV, Fasc. I. Platyhelminthes, Mésozoaires, Acanthocéphales, Némertiens.* Masson, Paris.

*Götte, A. 1902. *Lehrbuch der Zoologie.*

Gould, S.J. 1977. *Ontogeny and Phylogeny.* Harvard University Press, Cambridge, Massachusetts.

Gould-Somero, M. 1975. Echiura. Pp. 277–311 in Giese, A.C., and Pearse, J.S. (eds.), *Reproduction in Marine Invertebrates,* Vol. 3. Academic Press, New York.

Gray, J. 1931. *A Textbook of Experimental Cytology.* Cambridge University Press, Cambridge.

Guinot, D. 1978. Principes d'une classification évolutive des Crustacés Décapodes Brachyoures. *Bulletin Biologique de la France et de la Belgique 112:* 211–292.

Gurney, R. 1942. *Larvae of Decapod Crustacea.* Ray Society, London.

†Haeckel, E. 1874. *Anthropogenie oder Entwicklungsgeschichte des Menschen.* Leipzig.

Hall, B.K. 1984. Development mechanisms underlying the formation of atavisms. *Biological Reviews 59:* 89–124.

Harmer, S.F. 1887. Appendix. Pp. 39–47 in McIntosh, W. Report on *Cephalodiscus dodecalophus* McIntosh, a new type of the Polyzoa, procured on the voyage of H.M.S. Challenger during the years 1873–76. *Report on the Scientific Results of the Exploring Voyage of H.M.S. Challenger 1873–76. Zoology* 20(3), (part 62).

Horstadius, S. 1973. *Experimental Embryology of the Echinoderms.* Oxford University Press, Oxford.

*Huxley, T.H. 1875. The classification of animals. *Quarterly Journal of Microscopical Science 15.*

Hyman, L.H. 1940–59. *The Invertebrates,* Vols. I–V. McGraw-Hill, New York.

ICBN. 1972. *International Code of Botanical Nomenclature*. A. Oosthoek's Uiteversmaatschappij N. V., Utrecht.

ICZN. 1985. *International Code of Zoological Nomenclature*. International Trust for Zoological Nomenclature, London.

Imms, A.D. 1946. *A General Textbook of Entomology, Including the Anatomy, Physiology, Development and Classification of Insects*, 6th ed. Methuen, London.

Jablonski, D., and Lutz, R.A. 1983. Larval ecology of benthic marine invertebrates: Paleontological implications. *Biological Reviews 58:* 21–89.

Jägersten, G. 1972. *Evolution of the Metazoan Life Cycle. A Comprehensive Theory*. Academic Press, London.

Jefferies, R.P.S. 1979. The origin of chordates—a methodological essay. Pp. 443–477 in House, M.R. (ed.), *The Origin of Major Invertebrate Groups*. Systematics Association Special Volume 12. Academic Press, London.

Jefferies, R.P.S. 1986. *The Ancestry of the Vertebrates*. British Museum (Natural History), London.

Kerkut, G.A. 1960. *Implications of Evolution*. Pergamon Press, Oxford.

Kessel, M.M. 1964. Reproduction and larval development in *Acmaea testudinalis*. *Biological Bulletin 127:* 294–303.

Kirk, H.B. 1916. Much abbreviated development of a sand-star *(Ophionereis schayeri?)*. *Transactions of the New Zealand Institute 48:* 12–18.

Lacalli, T.C. 1984. The nervous system and ciliary bands of Müller's larva. *Proceedings of the Royal Society of London Series B 217:* 37–58.

Lemche, H. 1959. Protostomian relationships in the light of *Neopilina*. *Proceedings of the 15th International Congress in Zoology, London:* 381–389.

*Leuckart, R. 1854. Bericht über die Leistungen in der Naturgeschichte der niederen Thiere während der Jahre 1848–1853. *Archiv für Naturgeschichte 20(2)*.

Lewin, R. 1982. Can genes jump between eukaryotic species? *Science, New York 117:* 42–43.

†Lohmann, H. 1933. Tunicata. Pp. 1–202 in Kükenthal and Krumbach (eds.), *Handbuch der Zoologie*, Vol. 5, Part 2.

MacBride, E.W. 1911. Two abnormal plutei of *Echinus*, and the light they throw on the factors in the normal development of *Echinus*. *Quarterly Journal of Microscopical Science 57:* 235–250, 2 pls.

MacBride, E.W. 1914: *Text-Book of Embryology*. Vol. 1, *Invertebrata*. Macmillan, London.

Margulis, L. 1970. *Origin of Eukaryotic Cells*. Yale University Press, New Haven.

Margulis, L. 1981. *Symbosis in Cell Evolution*. Freeman, San Francisco.

Marshall, C.R. 1988. DNA–DNA hybridization, the fossil record, phylogenetic reconstruction, and the evolution of the clypeasteroid echinoids. Pp. 107–119 in Paul, C.R.C., and Smith, A.B. (eds.), *Echinoderm Phylogeny and Evolutionary Biology*. Clarendon Press, Oxford.

Matsumura, T., Masegawa, M., and Shigei, M. 1979. Colagen biochemistry and phylogeny of echinoderms. *Comparative Biochemistry and Physiology 62B:* 101–105.

Metschnikoff, E. 1881. über die systematische Stellung von *Balanoglossus*. *Zoologischer Anzeiger 4.*

Monod, T. 1956. Hippidea et Brachyura ouest-africaines. *Mémoires de l'Institut Français d'Afrique Noire 45.* I.F.A.N., Dakar.

Moore, A.R. 1957. Biparental inheritance in the interordinal cross of seaurchin and sand dollar. *Journal of Experimental Zoology 135:* 75–83.

Mortensen, T. 1921. *Studies of the Development and Larval Forms of Echinoderms*. G.E.C.Gad, Copenhagen.

Mortensen, T. 1931. Contributions to the study of the development and larval forms of echinoderms I-II. *Det Konglige Danske Videnskabernes Selskabs Skrifter, Naturvidenskabelig og Mathematisk* Afdeling 4, Raekke 4(1): 1–39, pls. I–VII.

Mortensen, T. 1937. Contributions to the study of the development and larval forms of echinoderms. III. *Det Konglige Danske Videnskabernes Selskabs Skrifter, Naturvidenskabelig og Mathematisk* Afdeling 9, Raekke 7(1): 1–65, pls. I–XV.

Mortensen, T. 1938. Contributions to the study of the development and larval forms of echinoderms. IV. *Det Konglige Danske Videnskabernes Selskabs Skrifter, Naturvidenskabelig og Mathematisk* Afdeling 9, Raekke 7(3): 1–33, pls. I–XII.

Müller, A.H. 1970. über den Sexualdimorphismus regulärer Echinoidea (Echinodermata) der Oberkreide. *Deutsche Akademie der Wissenschaften zu Berlin, Monatsberichte 12:* 923–935.

Paul, C.R.C. 1979. Early echinoderm radiation. Pp. 443–481 in House, M.R. (ed.), *The Origin of Major Invertebrate Groups*. Systematics Association Special Volume 12. Academic Press, London.

Paul, C.R.C., and Smith, A.B. 1984. Early radiation and phylogeny of echinoderms. *Biological Reviews of the Cambridge Philosophical Society 59:* 443–481.

Paul, C.R.C., and Smith, A.B. (eds) 1988. *Echinoderm Phylogeny and Evolutionary Biology*. Clarendon Press, Oxford.

Pilger, J. 1978. Settlement and metamorphosis in the Echiura: A review. Pp. 103–112 in Chia, F.S., and Rice, M.E. (eds.), *Settlement and Metamorphosis in Marine Invertebrate Larvae*. Elsevier, New York.

Pruho, H. 1888. Recherches sur le *Dorocidaris papillata. Archives de Zoologie Experimental et Géneral,* Sér. 2, Vol. 5.

Raff, R.A., Field, K.G., Ghiselin, M.T., Lane, D.J., Olsen, G.J., Pace, N.R., Parks, A.L., Parr, B.A., and Raff, E.C. 1988. Molecular analysis of distant phylogenetic relationships in echinoderms. Pp. 29–41 in Paul, C.R.C., and Smith, A.B. (eds.), *Echinoderm Phylogeny and Evolutionary Biology*. Clarendon Press, Oxford.

Rice, A.L. 1980. Crab zoeal morphology and its bearing on the classification of the Brachyura. *Transactions of the Zoological Society of London 35*: 271–424.

Rice, A.L., and von Levetzow, K.G. 1967. Larvae of *Homola* (Crustacea: Dromiacea) from South Africa. *Journal of Natural History 1*: 435–453.

Rice, A.L., and Provenzano, A.J. 1970. The larval stages of *Homola barbata* (Fabricius) (Crustacea, Decapoda, Homolidae) reared in the laboratory. *Bulletin of Marine Science 20*: 446–471.

Rice, M.E. 1975. Sipuncula. Pp. 67–127 in Giese, A.C., and Pearse, J.S. (eds.), *Reproduction of Marine Invertebrates*, Vol. 2. Academic Press, New York.

Rothwell, N.V. 1983. *Understanding Genetics*, 3rd ed. Oxford University Press, New York.

Salvini-Plawen, L. v. 1980. Was ist eine Trochophora? Eine Analyse der Larventypen mariner Prostomier. *Zoologischer Jahrbücher. Anatomie und Ontogenie der Tiere 103*: 389–423.

Salvini-Plawen, L. v. 1985. Early evolution and the primitive groups. Pp. 59–150 in Trueman, E.R., and Clarke, M.R. (eds.), *The Mollusca*, Vol. 10, *Evolution*. Academic Press, Orlando, Florida.

Sedgwick, A. 1898. *A Student's Text Book of Zoology*. Swan Sonnenschein, London.

Shearer, C., de Morgan, W., and Fuchs, H.M. 1914. On the experimental hybridization of echinoids. *Philosophical Transactions of the Royal Society of London Series B 204*: 255–362, 7 pls.

Shipley, A.E. 1896. Gephyrea and Phoronis. Pp. 411–462 in Harmer, S.F. and Shipley, A.E. (eds.) *The Cambridge Natural History*, Vol. 2. Macmillan, London.

Smiley, S. 1986. Metamorphosis of *Stichopus californicus* and its phylogenetic implications. *Biological Bulletin, Marine Biological Laboratory, Woods Hole, Massachusetts 171*: 611–631.

Smiley, S. 1988. The phylogenetic relationships of holothurians: A cladistic analysis of the extant echinoderm classes. Pp. 69–84 in Paul, C.R.C., and Smith, A.B. (eds.), *Echinoderm Phylogeny and Evolutionary Biology*. Clarendon Press, Oxford.

Smith, A.B. 1984. Classification of the Echinodermata. *Palaeontology 27*: 431–459.

Smith, A.B. 1988. Fossil evidence for the relationships of extant echinoderm classes and their times of divergence. Pp. 85–97 in Paul C.R.C., and Smith, A.B. (eds.), *Echinoderm Phylogeny and Evolutionary Biology*. Clarendon Press, Oxford.

Stace, C.A. 1980. *Plant Taxonomy and Biosystematics*. Arnold, London.

Strathmann, R.R. 1978. Larval settlement in echinoderms. Pp. 235–246 in Chia, F.S., and Rice, M. (eds.), *Settlement and Metamorphosis in Marine Invertebrate Larvae*. Elsevier, New York.

Syvanen, M.G. 1985. Cross-species gene transfer: Implications for a new theory of evolution. *Journal of Theoretical Biology 112:* 333–343.

Tattersall, W.M., and Sheppard, E.M. 1934. Observations on the asteroid genus *Luidia*. Pp. 35–61 in *James Johnstone Memorial Volume*. Liverpool University Press, Liverpool.

Thompson, T.E. 1961. The development of *Neomenia carinata* Tullberg. *Proceedings of the Royal Society of London Series B 153:* 263–278.

Thomson, C.W. 1874. *The Depths of the Sea*. Macmillan, London.

Tregouboff, G., and Rose, M. 1957. *Manuel de Planctonologie Méditerranéen*. Centre National de la Recherche Scientifique, Paris.

Verdonk, N.H., and Biggelaar, J.A.M. van den. 1983. Early development and the formation of the germ layers. Pp. 91–122 in Verdonk, N.H., Biggelaar, J.A.M. van den, and Tampa, A.S. (eds.), *The Mollusca. Vol. 3. Development*. Academic Press, Orlando, Florida.

Whittington, H.B. 1979. Early arthropods, their appendages and relationships. Pp. 253–268 in House, M.R. (ed.), *The Origin of Major Invertebrate Groups*. Systematis Association Special Volume 12. Academic Press, London.

†Willey, A. 1893. Studies on the Protochordata. No. 1. On the origin of the branchial stigmata, preoral lobe, endostyle, atrial cavities, etc. in *Ciona intestinalis* L. with remarks on *Clavellina lepadiformis*. *Quarterly Journal of Microscopical Science 34:* 317–360.

Williamson, D.I. 1960. A remarkable zoea, attributed to the Majidae (Decapoda, Brachyura). *Annals and Magazine of Natural History* (13)*3:* 141–144.

Williamson, D.I. 1976. Larval characters and the origin of crabs (Crustacea, Decapoda, Brachyura). *Thalassia Jugoslavica 10:* 401–414.

Williamson, D.I. 1982. The larval characters of *Dorhynchus thomsoni* Thomson (Crustacea, Brachyura, Majoidea) and their evolution. *Journal of Natural History 16:* 727–744.

Williamson, D.I. 1988a. Incongruous larvae and the origin of some invertebrate life-histories. *Progress in Oceanography 19:* 87–116.

Williamson, D.I. 1988b. Evolutionary trends in larval form. Pp. 11–25 in Fincham, A.A., and Rainbow, P.S. (eds.), *Aspects of Decapod Crustacean Biology (Symposia of the Zoological Society of London 59)*. Oxford University Press, Oxford.

Williamson, D.I. 1991. Sequential chimeras. Pp. 299–336 in Tauber, A.I.

(ed.), *Organism and the Origins of Self*. Kluwer, Dordrecht, Netherlands.

Wilson, D.P. 1932. On the mitraria larva of *Owenia fusiformis* Delle Chiaje. *Philosophical Transactions of the Royal Society of London Series B* 221: 231–334.

Wold, B.J., Klein, W.H., Hughes-Evans, B.R., Britten, R.J., and Davidson, E.H. 1978. Sea-urchin embryo mRNA sequences expressed in the nuclear RNA of adult tissues. *Cell* 14(4): 941–950.

Wright, A.D. 1979. Brachiopod radiation. Pp. 235–252 in House, M.R. (ed.), *The Origin of Major Invertebrate Groups*. Systematis Association Special Volume 12. Academic Press, London.

Yonge, C.M., and Thompson, T.E. 1976. *Living Marine Molluscs*. Collins, London.

Zimmer, R.L., and Woollacott, R.M. 1977. Metamorphosis, ancestrulae and coloniality in bryozoan life cycles. Pp. 91–142 in Woollacott, R.M., and Zimmer, R.L. (eds.), *Biology of Bryozoans*. Academic Press, New York.

Glossary

Annelida, annelids: a phylum (q.v.) of bilaterally symmetrical, segmented, coelomate worms, often with chitinous bristles but without a chitinous exoskeleton. The group includes earthworms, leeches, and polychaetes (q.v.).

Anomura, anomurans: a group of decapod crustaceans (q.v.), which includes the hermit crabs, squat lobsters, and porcelain crabs. The abdomen is protected with a mollusc shell or other "borrowed" covering or by holding it under the thorax; it is never as reduced as in the Brachyura (q.v.). Females without spermathecae. A typical zoea (q.v.) larva is shown in Figure 3.2e.

Archenteron: the central cavity in a gastrula (q.v.), lined with endoderm (q.v.). The inner end is blind, and the opening to the exterior is termed the blastopore. (See Fig. 5.2b.)

Asteromorpha, asteromorphs: a class (see "classification") of the Echinodermata (q.v.) comprising the starfishes or sea stars. The flattened body grades into five or more arms (see Fig. 5.1a). A bipinnaria larval stage (q.v.) is typical, often followed by a brachiolaria stage. Previously known as Asteroidea or asteroids.

Auricularia: a larval form occurring in the development of many Holothuromorpha (q.v.). There is no skeleton, and it is propelled by a continuous, convoluted band of cilia that encircles the mouth (Fig. 5.3d). The mouth develops as a deuterostome (q.v.), and the coelom (q.v.) is an enterocoel. It is succeeded by a doliolaria (q.v.).

Bipinnaria: a larval form occurring in the development of many Asteromorpha (q.v.). There are projecting arms with bands of locomotory cilia but no supporting skeleton; the ciliary bands usually form a preoral and a postoral loop (Figs. 5.3a, 7.2, 9.3c). The mouth develops as a deuterostome (q.v.) and the coelom (q.v.) is an enterocoel. The late bipinnaria may develop additional short arms with organs of attachment and is then known as a brachiolaria.

Bivalvia, bivalves: a class (see "classification") of the Mollusca (q.v.), which includes cockles, mussels, oysters, scallops, and clams. All have hinged, bivalved shells.

Blastopore: see "archenteron."

Blastula: an early stage in the development of triploblastic animals, consisting of a one-layered sphere of cells that may be hollow or contain mesenchyme cells or yolk cells. (See Fig. 5.2a.)

Brachiolaria: see "bipinnaria."

Brachyura, brachyurans: a group of decapod crustaceans (q.v.) comprising the true crabs. The abdomen is reduced and held under the thorax. Females have spermathecae (q.v.). A typical zoea (q.v.) larva is shown in Figure 3.2f.

Bryozoa, bryozoans: a phylum (see "classification") of bilaterally symmetrical, sessile, colonial, coelomate animals, with zooids of several types. Each feeding zooid possesses a lophophore (q.v.), and the anus opens outside this organ (Fig. 11.1g). Planktonic larvae occur in some species, and may be shelled, as in the cyphonutes (q.v.), or unshelled (Fig. 11.1e).

Chordata, chordates: a phylum (see "classification"), typically with a dorsal nerve chord and notochord, or the notochord may be replaced by a jointed column. Includes man and other vertebrates.

Cirripedia, cirripedes: barnacles and related crustaceans (q.v.). In the better known examples, the adult is sessile, with the body partly or wholly enclosed in a system of calcareous plates that are not moulted. Some are parasites and greatly modified. Hatch as nauplius larvae (Fig. 3.2a,b).

Classification: the grouping of individual organisms into species, species into genera, etc. The most commonly employed groupings are species (pl. species), genus (pl. genera), family, order, class, phylum (pl. phyla), and kingdom. Intermediate groups may be designated by using the prefixes sub- and super-, e.g., subspecies, superfamily.

Coelom: body cavity. One or more fluid-filled spaces in the mesoderm (q.v.) of many animals. It may be enterocoelic, i.e., formed from pouches that bud off from the archenteron (q.v. and Fig. 5.2c, d), or schizocoelic, i.e., formed from splits in the mesoderm.

Concentricyclomorpha, concentricyclomorphs: a class (see "classification") of the Echinodermata (q.v.) known as sea daisies. The single known species consists of a disc-like body with a ring of petal-like spines round the circumference, without arms and alimentary canal (Fig. 5.1f). Development is direct (q.v.). Previously known as Concentricycloidea.

Crinomorpha, crinomorphs: a class (see "classification") of the Echinodermata (q.v.) comprising the feather stars and sea lilies. The mouth points upward, there are flexible arms, often numerous and branched, and an aboral stalk is present for at least part of the life history (Fig. 5.1e). The free larva, when present, is a nonfeeding doliolaria (q.v.). Previously known as Crinoidea or crinoids.

Cross-fertilization: the penetration of an egg of one species by a sperm of another, followed by fusion of the two nuclei and then by cell division.

Crustacea, crustaceans: the group of animals, variously rated as a phylum, subphylum, or class (see "classification"), which includes shrimps, lobsters, and crabs, typically with a jointed external skeleton, made largely of chitin, that has to be moulted at intervals to permit growth, and with two pairs of antennae and one pair of mandibles.

Cyphonautes: a ciliated larva with a triangular, bivalved shell (Fig. 11.1f), occurring in the life histories of some Bryozoa (q.v.).

Decapoda, decapods: a group of crustaceans (q.v.) that contains the crabs, lobsters, and shrimps. The thoracic appendages consist of three pairs of maxillipeds and five pairs of legs. The last pair of abdominal appendages, the uropods, develops independently of and usually before the other abdominal appendages, the pleopods.

Desor's larva: a stage in the development of some Nemertea (q.v.), resembling a pilidium (q.v.) but occurring within the egg membrane (i.e., not free-living).

Deuterostome: a mouth that develops as a new aperture, independent of the blastopore (see "archenteron"), or an animal with such a mouth.

Direct development: development without a larval stage. The form that hatches resembles a miniature adult.

DNA: dioxyribonucleic acid, the chemical that makes up the genes of plants, animals, and other organisms.

Doliolaria: a barrel-shaped larva encircled by three to five bands of cilia, occurring in the development of some Holothuromorpha (q.v.), Crinomorpha (q.v.), and Ophiuromorpha (q.v.). The mouth develops as a deuterostome (q.v.) and the coelom (q.v.) is an enterocoel (q.v.). Examples shown in Figures 5.3e and 8.2e.

Echinodermata, echinoderms: a phylum (q.v.) of marine animals whose living representatives include sea-urchins, starfish (sea stars), brittle-stars, sea cucumbers, and sea lilies (feather stars). The adults all show radial symmetry, and the dermis produces spines and other calcareous structures. Examples of adults are shown in Figure 5.1, larvae in Figure 5.3.

Echinomorpha, echinomorphs: a class (see "classification") of the phylum Echinodermata (q.v.) comprising the sea-urchins, heart urchins, and sand dollars. The adults do not have free arms (Fig. 5.1c). They typically develop through a pluteus larval stage (q.v.). Previously known as Echinoidea or echinoids.

Echiura, echiurans: a phylum (q.v.) of bilaterally symmetrical, unsegmented, vermiform animals. There is a proboscis anterior to the mouth, and a convoluted gut leads to a posterior rectum (Fig. 10.1b). Usually hatch as a trochophore (q.v.).

Ectoderm: the outer layer of cells of a multicelled animal and organs derived from this layer.

Embryo: developing animal in an unborn or unhatched state, within the egg membrane.

Endoderm: the inner layer of cells of a multicelled animal, forming most of the gut and associated organs.

Enterocoel: see "coelom."

Enteropneusta, enteropneusts: a class (q.v.) of the Hemichordata (q.v.) with a long proboscis, no lophophoral arms, and many gill slits (Fig. 6.1a). Usually with a tornaria larva (q.v. and Fig. 6.1c).

Eukaryotes: single- or multicelled organisms [including protoctsts (q.v.), fungi, animals, and plants] in which each cell-nucleus is enclosed in a membrane (compare prokaryotes).

Evolution: literally unrolling, hence development. Organic evolution is the doctrine that all living things originated by descent, with modifications, from preexisting forms.

Gastropoda, gastropods: a class (see "classification") of the Mollusca (q.v.), which includes the snails and slugs; typically with a coiled shell, but this may be modified of absent.

Gastrula: stage in the early development of triploblastic animals immediately following the blastula (q.v.), consisting essentially of two layers of cells, the inner of which encloses the archenteron (q.v.). (See Fig. 5.2b.)

Gene: the unit of heredity; part of a nucleic acid molecule, containing coded information that can produce or contribute to one or more heritable features of the organism.

Genome: the total genetic complement of an organism.

Götte's larva: a larva resembling Müller's larva (q.v.) but with only four ciliated lobes.

Hemichordata, hemichordates: a phylum (q.v.) of bilaterally symmetrical animals with the body in three regions, each with one or two coelomic cavities. Pharynx with one or many slits to exterior. Adults and larvae shown in Figure 6.1. Sometimes known as Stomochordata.

Heterochrony: evolutionary changes in the rate of development, so that features that previously occurred only in the larva may be retained in the adult (paedomorphosis, neoteny, or progenesis) or features that previously occurred only in the adult may appear in the larva (acceleration or adultation).

Heterozygote: an egg cell of one species fertilized by a sperm of another.

Histogenesis: the formation and differentiation of tissues.

Histolysis: the breakdown of tissues, in which the cells lose their special shapes and functions.

Holothuromorpha, holothuromorphs, holothurians: a class (see "classification") of the Echinodermata (q.v.) comprising the sea cucumbers.

The body tends to be elongated and shows some degree of bilateral symmetry; the body wall is leathery (Fig. 5.1d). Typically develop through an auricularia (q.v.) stage, followed by a doliolaria (q.v.). Previously known as Holothuroidea or holothuroids.

Juvenile: an immature phase in the development of an animal, resembling a miniature adult in most respects. (Compare "larva.")

Kingdom: a major group of organisms. The five kingdoms recognized by Margulis (1981) are Monera (q.v.), Protoctista (q.v.), Fungi, Animalia, and Plantae.

Larva: an immature phase, differing considerably from the adult, in the postembryonic development of an animal. (Compare "embryo" and "juvenile.")

Lophophore: a crescentic, tentaculated, ciliated organ which carries food particles to the mouth in several groups of animals, including pterobranchs (q.v.) (Fig. 6.1b) and bryozoans (q.v.) (Fig. 11.1g).

Megalopa: a larval phase (see "larva") in the development of a crustacean (q.v.) in which the abdominal pleopods are functional. Usually follows a zoeal phase (see "zoea").

Mesenchyme: diffuse connective tissue cells in a jelly-like matrix.

Mesoderm: cells between the ectoderm and endoderm (q.v.) of many animals, which line the coelom (q.v.) and give rise to a number of organs.

Metamerism: serial repetition of organs.

Metamorphosis: change of form during development, as from a larva to an adult or from one larval phase to another.

Metatroch: a ring of cilia encircling a larva posterior to the mouth.

Metatrochophore: a late trochophore larva (q.v.) with developing tissues and organs of the next phase in development.

Metazoa, metazoans: multicelled animals made up of more than one type of cell.

Molecular biology: the study of the chemical molecules that carry coded genetic information.

Mollusca, molluscs, or mollusks: a phylum (q.v.) of bilaterally symmetrical animals including clams, snails, slugs, and octopuses, unsegmented or with limited metamerism, without a coelom other than that provided by the blood sinuses, frequently with a shell for protection and a foot for locomotion. Some hatch as trochophores (q.v.).

Monera: prokaryotic organisms (bacteria) comprising one of the five kingdoms (q.v.).

Müller's larva: a larva occurring in the development of many Polycladida (q.v.) in which the main natatory cilia are in a preoral band that extends into 8 or 10 lobes (Fig. 11.1a). The mouth develops from part of the blastopore (see "archenteron") and is therefore a protostome (q.v.).

Nauplius: a larval crustacean (q.v.) with three pairs of functional appendages, which will eventually become the two pairs of antennae and the mandibles of the adult. There is typically a small, median eye. Examples shown in Figure 3.2a–c.

Nemertea, nemertines (also known as **Nemertina, Nemertinea, Nemertini,** and **Rhynchocoela**): a phylum (q.v.) of bilaterally symmetrical worms, without a body cavity, with a proboscis. Serial repetition of some organs but body wall not segmented (Fig. 11.1d). Some with a pilidium larva (q.v.) or a Desor's larva (q.v.).

Neoteny: see "heterochrony."

Nucleotides: repeated units that together make up molecules of DNA (q.v.) and related compounds.

Ontogeny: the development of an organism through the various phases of its life history.

Ophiuromorpha, ophiuromorphs: a class (see "classification") of Echinodermata (q.v.) comprising the brittle-stars. The flexible arms are sharply demarcated from the disc-shaped body. The mouth faces downward and there is no intestine or anus (Fig. 5.1b). Development may be direct (q.v.), via a pluteus larva (q.v.) or (exceptionally) via a doliolaria larva (q.v.). Previously known as Ophiuroidea or ophiuroids.

Paedomorphosis: see "heterochrony."

Pelagosphaera: a larval phase, usually preceded by a trochophore (q.v.), in the development of most Sipuncula (q.v.), with a coelom, anterior mouth, and dorsal anus, usually propelled by a metatroch (q.v.) (Fig. 10.3f).

Pericalymma: another name for a test cell larva (q.v.).

Phylogeny: the history of group of organisms, including its genealogy and relationships.

Phylum: a major group of monerans (q.v.), protoctotists (q.v.), fungi, animals, or plants. (See also "classification.")

Pilidium: a helmet-shaped, ciliated larva occurring in the life history of some Nemertea (q.v.). The juvenile develops as a quasiparasite within the larva (Fig. 11.1c).

Plankton: organisms that live suspended in the water, with feeble swimming powers or none.

Platyhelminthes, platyhelminths: a phylum (q.v.) of animals comprising the flatworms. They are bilaterally symmetrical and unsegmented. The body consists of three layers, ectoderm (q.v.), endoderm (q.v.), and mesenchyme (q.v.), but there is no coelom (q.v.).

Pluteus: a form of larva occurring in the Echinomorpha (q.v.) and Ophiuromorpha (q.v.) with ciliated arms supported by calcarious rods. The larvae are deuterostomatous (q.v.) and enterocoelous (see "coelom"). Examples shown in Figures 5.3b,c, 7.1, 8.1, 8.2a–d, 9.3a,b.

Polychaeta, polychaetes: marine Annelida (q.v.) with many bristles (chaetae) arising from paired segmental prominences (parapodia). Many hatch as trochophores (q.v.).

Polycladida, polyclads: an order (see "classification") of Turbellaria (q.v.) in which the gut extends into many lobes (Fig. 11.1b). Usually develop through a Müller's larva (q.v.) or Götte's larva (q.v.).

Progenesis: see "heterochrony."

Prokaryotes: organisms, such as bacteria and viruses, in which there is no membrane-bound cell-nucleus (compare *eukaryotes*).

Protoctista: one of the five kingdoms (q.v.) of organisms, comprising unicellular and colonial unicellular organisms and multicellular algae.

Protostome: a mouth that develops directly from the blastopore (see "archenteron"), or an animal with such a mouth.

Prototroch: a ring of cilia encircling a larva anterior to the mouth.

Pterobranchia, pterobranchs: a class (see "classification") of the phylum Hemichordata (q.v.) with a shield-shaped proboscis, one to nine pairs of lophophoral arms, and only one gill slit.

Rhynchocoela: see Nemertea.

Schizocoel: see "coelom."

Segmentation: division of the body into semiindependent, serially repeated units (segments).

Sipuncula, sipunculans: a phylum (q.v.) of bilaterally symmetrical, vermiform, unsegmented, coelomate animals. The mouth leads to an extensible "introvert." The anus opens on the anterior dorsal body wall. Most pass through two larval phases: a trochophore (q.v.) followed by a pelagosphaera (q.v.).

Species: a group of organisms that potentially can interbreed. (See also "classification.")

Spermatheca: a sac in which a female may store sperm for the subsequent fertilization of her eggs.

Stomochordata: an alternative name for the Hemichordata (q.v.).

Syncytium: a structure with many cell nuclei but with no boundaries between the cells.

Taxonomy: the classification of organisms and the study of the principles involved.

Telson: the terminal part of a crustacean (q.v.), sometimes fused with the last abdominal segment, sometimes forming a separate segment.

Test cell larva: a larval form of some molluscs (q.v.) propelled by one or more transverse bands of cilia and having a single external aperture, the "pseudoblastopore," which opens posteriorly. The larval ectoderm consists of very large cells and is shed at metamorphosis.

Tornaria: larva occurring in the development of the Enteropneusta (q.v.), propelled by a convoluted band of cilia that encircles the mouth (Fig. 6.1c). It is a deuterostome (q.v.) with an enterocoel (see "coelom").

Trimerous: having three parts. (Applied to animals such as insects and hemichordates.)

Triploblastic: having three layers of cells, usually ectoderm, endoderm, and mesoderm, which originate early in the development of an animal.

Trochophore, trochosphere: a larval form that follows the gastrula (q.v.) in the development of animals of many phyla. The mouth develops from part of the blastopore (see "archenteron"), and cilia are grouped in a preoral ring (prototroch) and usually a tuft at each end (Figs. 10.2a, 10.3a,b,d,e, 10.4a).

Turbellaria, turbellarians: a class (see "classification") of the Platyhelminthes (q.v.), mostly free-living, ciliated worms, each with a mouth and pharynx but no anus.

Urochordata, urochordates: usually considered to be a subphylum of the phylum Chordata (q.v.), consisting of ascidians (Fig. 13.1), thaliaceans, and appendicularians. Most ascidians and some thaliaceans have tadpole larva; the appendicularians mature as tadpoles. See Chapter 13 for suggested revision.

Veliger: a shelled larva of the Mollusca (q.v.), with natatory cilia on a special organ, the velum, which usually extends into lobes (Fig. 10.4b,c).

Zoea: a larval crustacean (q.v.) that swims with the outer branches of its biramous thoracic appendages, a varying number of which may be functional. Paired, compound eyes are usually present (Figs. 3.2d–f, 4.1b,d,f).

Zygote: a fertilized egg cell.

Index

electrophoresis, 160, 162–164
embryo, embryonic, embryology, 1, 4,
 9, 10, 13, 20, 27, 43, 52, 58, 60, 71,
 72, 74 (Fig. 7.4a–d), 77, 80, 81, 91,
 94, 96, 108, 115, 128, 129, 149, 150,
 152–154, 157, 158, 160, 161, 167, 168,
 187, 190, 192–194, 196–198, 209
endoderm, 12, 63, 64, 128, 210
Endoprocta, 10, 140
enterocoel, enterocoelous, enterocoely,
 11, 44, 52, 57, 64, 71, 73, 76, 108,
 117, 128, 139, 149, 150, 157, 165, 173,
 196, 210
Enteropneusta, enteropneusts, 3, 11,
 52–54 (inc. Fig. 6.1a, c–f), 57, 59,
 61, 63, 64, 103, 128, 146, 151, 155,
 163, 165, 167, 191, 210
enzymes, 163, 164
Eocrinomorpha, 98 (Fig. 9.1a)
eukaryotes, 15, 210
evolution, evolutionary, evolve, xi, 1,
 5, 6, 14, 15, 19, 37, 48–50, 58, 59, 65,
 76, 77, 79–81, 87, 94, 95, 103, 115,
 116, 117, 120, 128, 129, 130, 136, 140,
 142, 145–148, 150, 151, 160, 162, 187,
 188, 191, 192, 195, 196, 198, 210
 clandestine, 81, 82, 89
 convergent, 2, 80, 81, 86, 87, 108,
  114, 134, 145, 162, 163
 divergent, 79, 115, 147
feather stars, 43
Fell, H.B., 50, 58, 71–73, 74 (Fig. 7.4),
 81, 82, 86, 92 (Fig. 8.2), 104 (Fig. 93),
 149, 200
Fernando, W., 128, 200
fertilization, fertilize, 135, 138, 145,
 153, 155, 156, 158, 159, 166, 171–174,
 176, 178 (Fig. 13.3), 181 (Fig. 13.4),
 182, 187, 193, 195
flatworms, 3, 18 (Fig. 3.1), 131, 135,
 141, 167, 196 (see also
 Platyhelminthes)
Flickinger, R.A., 172, 173, 200
fossils, fossil record (inc.
 palaeontology), 5, 33–35, 49, 87,
 96–102 (inc. Figs. 9.1, 9.2), 106, 107,
 109, 111–113, 115, 160, 161, 191
Frey, H., 52, 200
Frost, I., 196

Flustrellidra, 138

Gammarus, 184
Garstang, W., 134, 195, 200, 201
Gastropoda, gastropods, 28, 110 (Fig.
 10.1d), 113, 124, 126–129 (inc. Fig.
 10.4), 150, 210
gastrula, 10, 44–46 (inc. Fig. 5.2b), 64,
 70, 72, 73, 74 (Fig. 7.4b), 90, 113,
 175, 210
genes, genetic, 2, 5, 14–16, 20, 58, 59,
 70, 76, 79, 87–89, 94, 103, 106, 120,
 146, 150, 152–158, 161, 163, 166, 167,
 169, 188, 190–193, 197, 210
 genetic engineering, 14–16, 189
 genetic transfer, xi, 6, 16, 20, 50, 76,
  78, 80, 128, 129, 134, 141, 142, 145,
  148–153, 159, 162, 183, 189, 191,
  194
genital aperture, genital pore, 42, 96,
 106, 107, 162
genome, 15, 17, 170, 188–190, 197, 210
Gerould, 122 (Fig. 10.3)
Gibbs, P.E., 110 (Fig. 10.1), 201
Giudice, G., 89, 153, 171, 201
Goad, L.J., 88, 201
Gogia, 98 (Fig. 9.1f)
Goldschmidt, R., 192, 201
Golfingia, 110 (Fig. 10.1c), 111, 122
 (Fig. 10.3d)
Gontscharoff, M., 136, 137, 201
Gould, S.J., 1, 201
Gould-Somero, M., 120, 122 (Fig.
 10.3), 201
Gorgonocephalidae, 73, 149, 162, 165–
 167
Götte, A., 53, 201
Götte's larva, 11, 131, 136, 210
Gray, J., 156, 201
Guinot, D., 32, 33, 35, 201
Gurney, R., 30, 201
Gymnolaemata, 140

haploid, 153
Haeckel, E., 58, 195, 201
Haliotis, 110 (Fig. 10.1d), 113
Hall, B.K., 36, 206
Harmer, S.F., 52, 201
Hatschek, 122 (Fig. 10.3)

MacBride, E.W., 62 (Fig. 7.1), 67, 69
(Fig. 7.3), 202
macromutation, 15, 188
Macrosetella, 98 (Fig. 9.1g)
Marcus, 132 (Fig. 11.1)
Margulis, L., xi, 15, 188, 189, 198,
202
Marshall, C.R., 16, 203
Matsumura, T., 88, 203
megalopa, 13, 32, 34, 147, 211
Mellita, 42
Membranipora, 132 (Fig. 11.1f), 138
mesenchyme, 64, 66, 72, 73, 83, 135,
211
mesocoel, 61, 62 (Fig. 7.1), 63, 67, 70,
77, 78
mesoderm, 63, 117, 120, 121, 128, 137,
138, 211
metacoel, 61, 63, 77, 78
metamerism, metameric, 111, 113,
119–121, 192, 197, 211
metamorphose, metamorphosis, 2, 4–
6, 9, 12, 13, 46, 54 (Fig. 6.1c–f), 59–
61, 65–67, 77–79, 103, 108, 114, 117,
118 (Fig. 10.2), 119–124 (inc. Fig.
10.3), 125, 129, 134, 136, 141, 142,
147–149, 151, 154–157, 169–174,
176, 181–183 (inc. Fig. 13.4), 187,
192–195, 211 (*see also* cataclysmic
metamorphosis)
metatroch, 113, 114, 121, 211
metatrochophore, 114, 211
Metazoa, metazoans, 10, 18 (Fig. 3.1),
44, 56–58, 80, 100, 149, 211
Metschnikoff, E., 52, 203
Miocidaris, 102, 103
mitraria, 117, 118 (Fig. 10.2)
Mollusca, molluscs, 3, 12, 18 (Fig.
3.1), 21, 28, 39, 108, 110–113 (inc.
Fig. 10.1d), 114–116, 124–129 (inc.
Fig. 10.4), 134, 135, 141, 150, 152,
156, 157, 163, 167, 169, 196, 211
Monod, T., 30, 203
Monoplacophora, 113
monster, 187, 192, 193
Moore, A.R., 172, 173, 203
Mortensen, T., 47 (Fig. 5.3), 78, 83, 84
(Fig. 8.1), 90, 92 (Fig. 8.2), 104 (Fig.
9.3), 173, 203

moth, 149
mouth, 11, 37–39, 42–45 (inc. Fig.
5.2), 48, 56, 57, 63–65, 72, 73, 77,
109, 111–113, 119, 121, 129, 135,
137–140, 150, 157, 172
Müller, A.H., 107, 203
Müller, J., 132 (Fig. 11.1), 136
Mülleria, 134
Müller's larva, 11, 131, (Fig. 11.1), 132
(Fig. 11.1a), 134, 135, 136
mutation, 14, 15, 125, 188, 191, 211
myriopod, 66

natural selection, xi, 2, 14, 88, 120,
129, 148, 169, 198
nauplius, 13, 21, 22 (Fig. 3.2a–c), 116,
211
near-trochophore, near-trochophorate,
131, 140
nectochaete, 4, 12, 117, 129
Nemertea, nemertines, 3, 18 (Fig. 3.1),
131, 132 (Fig. 11.1c, d), 136–138,
142, 155, 163, 167, 171, 196, 212
Nemertes, 137
Nemertina, Nemertini, 137
neo-Darwinist, neo-Darwinism, neo-
Darwinian, 14, 119, 187, 188
Neomenia, 126 (Fig. 10.4d–i)
neoteny, 37, 49, 187, 212
Nereis, 110 (Fig. 10.1a)
Norton, T.A., 189
nuclear membrane, 154, 155
nucleotides, 166, 167, 212
nucleus, nuclear, 152–155, 160, 165,
166, 176

Obelia, 83
Oligochaeta, 109
ontogeny, 1, 58, 129, 130, 147, 148,
187, 212
onychophoran, 66
Ophiocomina, 61, 62 (Fig. 7.1)
Ophioderma, 91, 94
Ophiodermatidae, 91
Ophiolepidae, 91, 165
Ophiolepis, 91–94 (inc. Fig. 8.2e), 165
Ophiomyxidae, 73, 149, 162, 165–167
Ophiopluteus, 83, 84 (Fig. 8.1)
ophiopluteus, 46, 80–88, 91, 92 (Fig.
8.2c, d), 106

test-cell larva, 12, 114, 124, 125–128
(inc. Fig. 10.4d, e), 213
Thalassema, 120
Thaliacea, thaliaceans, 194, 196
Thompson, T.E., 125, 126 (Fig. 10.4), 205
Thomson, C.W., 30, 205
tornaria, 11, 54 (Fig. 6.1c), 57, 59–61, 66, 103, 146, 149–151, 165, 213
transfer of larval form, *see* larval transfer
tree of life, 17 (*see also* phylogenetic tree)
Tregouboff, G., 22 (Fig. 3.2), 205
Triassic, 96, 100, 102, 106
Tribrachidium, 96, 97
trimerous, 53, 56, 140, 213
triploblastic, 70, 213
triradial, *see* radial
Trochocystites, 98 (Fig. 9.1d)
trochophore, trochophorate, 3–5, 11–13, 108–110 (inc. Fig. 10.1), 113–131 (inc. Figs. 10.2a, 10.3a, b, d, e, 10.4a, b), 134–136, 138, 140, 141, 148, 149, 151, 152, 156, 163, 189, 192, 214
Turbellaria, turbellarians, 131, 132 (Fig. 11.1a, b), 135–137, 140, 141, 159, 214
Tunicata, 194
types of larvae: *see* larval types

Ubisch, 89, 94

Urechis, 120, 122 (Fig. 10.3c)
Urochordata, urochordates, 12, 174, 187, 194, 195, 214

veliger, 12, 13, 124–126 (inc. Fig. 10.4c), 129, 214
velum, 124
Verdonk, N.H., 152, 205
Vertebrata, vertebrates, 174, 195
virus, 156
Verril, 132 (Fig. 11.1)
Viviparus, 108, 128, 150, 157

Whittington, H.B., 116, 205
Willey, A., 195, 205
Williamson, D.I., 6, 29, 32, 33–36, 38, 87, 160, 189, 205
Wilson, D.P., 118 (Fig. 10.2), 119, 206
Wold, B.J., 152, 155, 206
worms, 109, 112, 120, 131, 136, 138, 148, 155, 167
Wright, A.D., 116, 206

Xiphosura, 116
Xyloplax, 40 (Fig. 5.1), 43, 73

Yonge, C. M., 112, 206

Zimmer, R.L., 139, 206
zoea, 13, 23 (Fig. 3.2d–f), 27, 28–35 (inc. Fig. 4.1b, d, f), 50, 147, 149, 158, 170, 214
zooid, 139
zoological nomenclature, 38